Advances in Intelligent and Soft Computing 113

Editor-in-Chief: J. Kacprzyk

Advances in Intelligent and Soft Computing

Editor-in-Chief

Prof. Janusz Kacprzyk
Systems Research Institute
Polish Academy of Sciences
ul. Newelska 6
01-447 Warsaw
Poland
E-mail: kacprzyk@ibspan.waw.pl

Further volumes of this series can be found on our homepage: springer.com

Jörn Altmann, Ulrike Baumöl,
and Bernd J. Krämer (Eds.)

Advances in Collective
Intelligence 2011

 Springer

Editors

Prof. Dr. Jörn Altmann
Seoul National University
College of Engineering
Department of Industrial Engineering
599 Gwanak-Ro, Gwanak-Gu
Seoul 151-742
South Korea
E-mail: jorn.altmann@acm.org

Prof. Dr. Ulrike Baumöl
FernUniversität in Hagen
Fakultät für Wirtschaftswissenschaft
Lehrstuhl für Betriebswirtschaftslehre,
insb. Informationsmanagement
Universitätsstrasse 41
58097 Hagen
Germany
E-mail: ulrike.baumoel@fernuni-hagen.de

Prof. Dr. Bernd J. Krämer
FernUniversität in Hagen
Fakultät für Mathematik und Informatik
Lehrstuhl für Datenverarbeitungstechnik
Universitätsstrasse 27
58097 Hagen
Germany
E-mail: bernd.kraemer@fernuni-hagen.de

ISBN 978-3-642-25320-1   e-ISBN 978-3-642-25321-8

DOI 10.1007/978-3-642-25321-8

Advances in Intelligent and Soft Computing   ISSN 1867-5662

Library of Congress Control Number: 2011940960

Typeset by Scientific Publishing Services Pvt. Ltd., Chennai, India

Printed on acid-free paper

5 4 3 2 1 0

springer.com

Preface

Welcome to the proceedings of the second Symposium on Collective Intelligence (COLLIN 2011), which was held in collaboration with SDPS 2011 on Jeju Island, South Korea. As collective intelligence is a truly transdisciplinary field of science with transformative potential, it is a perfect enlargement of SDPS's mission.

Collective intelligence has become an attractive subject of interest for both academia and industry. More and more conferences and workshops discuss the impact of the users' motivation to participate in the value creation process, the enabling role of leading-edge information and communication technologies and the need for better algorithms to deal with the growing amount of shared data. There are many interesting and challenging topics that need to be researched and discussed with respect to knowledge creation, creativity and innovation processes carried forward in the emerging communities of practice.

COLLIN is on the path to become the flagship conference in the areas of collective intelligence and ICT-enabled social networking. We were delighted to again receive contributions from different parts of the world including Australia, Europe, Asia, and the United States. Encouraged by the positive response, we plan COLLIN 2012 to be held next year end of August at FernUniverstität in Hagen.

In order to guarantee the quality of the event, each paper went through a double-blind review process. The reviews concentrated on originality, quality and relevance of the paper topic to the symposium. In addition, we invited a few renowned experts in the field to contribute to the success of the symposium with outstanding papers reporting on their most recent research. Our special thanks go to the authors for submitting their papers, to the international program committee members, and to numerous reviewers who did an excellent job in guaranteeing that the papers in this volume are of very high quality.

August 2011

Jörn Altmann
Ulrike Baumöl
Bernd J. Krämer

Organization

On the organization side, we are indebted to the organizational committees of COLLIN 2011 and SDPS 2011 for their generous, invaluable help and support in all aspects of the organization of the COLLIN track. Our special thanks also go to Alexander Kornrumpf who dedicated considerable time and effort to the final editing of this volume.

Conference and Program Chairs

Jörn Altmann
Seoul National University, Korea

Ulrike Baumöl
Fernuniversität in Hagen, Germany

Bernd Krämer
Fernuniversität in Hagen, Germany

Program Committee

Rajendra Akerkar
Vestlandsforsking - Western Norway Research Institute, Norway

Stuart Evans
Carnegie Mellon University, Campus West, USA

Kai Fischbach
Universität zu Köln, Germany

Sabine Fliess
FernUniversität in Hagen, Germany

Oliver Gassmann
University St. Gallen, Switzerland

Contents

Enough Questions for Everybody

Peter Miller

While I was writing my book, *The Smart Swarm*, about collective intelligence in nature and society, I often felt like one of the bees depicted on the front cover, buzzing from one field of research to another to pick up the latest thinking. As I worked my way through the widespread and expanding landscape, I met biologists, physicists, computer scientists, sociologists, engineers, psychologists, economists, political scientists, network theorists, and neuroscientists, and I began to see broad connections between the problems they were tackling. Biologists were talking about self-organization in superorganisms, while economists were debating the self-correcting tendencies of markets. Physicists were modeling collective motion, while psychologists were measuring collective biases in decision-making. Sociologists were exploring the wisdom of crowds, while engineers were experimenting with smart teams of robots. Running through all these discussions was a common thread that seemed obvious even to a non-scientist like me: Groups in nature have evolved ways to squeeze intelligence from relatively simple ingredients, and if we could just figure out how they do it we might learn something useful.

As many of you already know, colonies of ants, honeybees, and termites – as well as flocks of birds, schools of fish, herds of land animals – coordinate the efforts of thousands or even millions of individuals to accomplish feats that none of their members can handle alone, such as moving heavy objects, building elaborate nests, locating the best sources of food, or following migration routes. As they pick up information from one another and respond appropriately, members of a group form huge, mobile sensor nets to alert one another of dangers or opportunities. By following relatively simple rules and constantly interacting with one another, natural swarms take care of business in a timely, flexible manner. And they do so without bosses through various forms of self-organization. They're so good at it, in fact, they often make better decisions as groups than we do, despite our superior brainpower.

Because of their appeal as models of collective intelligence, ant colonies and other groups in nature have attracted the attention of researchers in many disciplines.

Peter Miller
National Geographic Magazine

J. Altmann et al. (Eds.): Advances in Collective Intelligence 2011, AISC 113, pp. 1–3.
springerlink.com   © Springer-Verlag Berlin Heidelberg 2012

Five years ago, when I first started attending conferences such as COLLIN 2011, I was struck by how interdisciplinary these events were. Biologists gave talks to engineers. Mathematicians gave talks to biologists. Physicists gave talks to economists. Sometimes the gulfs between the disciplines seemed too wide to bridge, as members of one tribe struggled to understand the secret language of another. At other times, sparks of imagination jumped around the room with an almost electric crackle. At such times, it became clear to me that the scientists making the biggest efforts to connect with other disciplines were the same ones doing the most creative research. As Deborah Gordon, a biologist at Stanford University put it, "I think it helps us all to talk to each other, not because it will turn out that ant colonies will work exactly like brains or gene interactions, but because it will stretch my way of thinking. What's most helpful for someone like me who works with complex systems is to keep learning about the details of other systems, because it helps us all to get better at thinking about these kinds of systems."

Occasionally the benefit is more direct. As Julia Parrish, a biologist at the University of Washington told me, a colleague of hers came back from an engineering conference with a new appreciation for the way that herding dogs get feedback from the animals they're controlling. "But he said he wouldn't have thought that way unless he had been working with engineers on control theory projects," she added. Similarly, Jeremiah Cohen, a neuroscientist at Harvard University, returned from a workshop on social insects impressed with similarities between the behavior of neurons involved in eye movement and worker ants and bees. Insect colonies, like neurons, seem to use quorum decision-making rules to process multiple signals from the environment and translate them into action. "Does the brain do the same thing to make sense of what we're seeing?" he wondered.

Even as researchers like these are finding inspiration in projects outside their fields, other scientists are striving to pull it all together under the vast umbrella of complex systems. For the past 25 years, multidisciplinary teams at the Santa Fe Institute in New Mexico have been tackling complex subjects such as metabolic networks, phase transitions, and financial markets. "Many of society's most interesting observations, and most pressing problems, fall far from the concerns of disciplinary research," the Institute says on its website. "Complex systems research attempts to uncover and understand the deep commonalities that link artificial, human, and natural systems."

In Europe, meanwhile another group of complexity scientists recently proposed a ten-year, billion-dollar effort called FutureICT aimed at, among other things, predicting how political decisions will impact the world's environment and economy. Using super computers, they ambitiously propose to model relevant natural and human systems to anticipate "imminent techno-socio-economic crises" such as climate change, environmental destruction, political conflicts, and financial crashes and then to come up with measures to alleviate or avoid them. "It's time to explore social life on Earth, and everything it relates to, in the same ambitious way that we have spent the last century or more exploring our physical world," stated Dirk Helbing, a former physicist turned sociologist at the Swiss Federal Institute of Technology in Zurich.

Alternatively, some experts predict it will be network science, now still in its early years, that eventually provides a common paradigm to study complex systems from honeybee colonies to electric power grids. "The role of networks in this area is obvious," writes Albert-László Barabási of Northeastern University. "All systems perceived to be complex, from the cell to the Internet and from social to economic systems, consist of an extraordinarily large number of components that interact via intricate networks." As we learn more about the structures of different networks we can see more clearly how they shape the flow of information, material, or energy passing through them – what they enable or suppress. Although a beehive's food collection system may not seem to have much in common with a power grid, they're both networks of interactions. What each bee does affects the behavior of all the other bees, just as each power plant affects the behavior of all the others on the grid.

Looking down the road at future research topics, the participants of one workshop I attended put together a kind of to-do list of questions. As one might expect, the questions were interdisciplinary in nature. Here's a sampling: How does variation among individuals in a group contribute to its collective intelligence? Can biases at the individual level actually lead to better decisions at the group level? What problems does variation present to coordination and cooperation? Many animals take part in "selfish herds" in which they benefit as a group from individuals acting in their own interests – could robots do that too? Does a superorganism learn the same way that an individual does? Can it filter out noise the same way? Can it adapt to new challenges? Can it behave strategically? Can it retain cultural information? How do groups in nature and human organizations differ when it comes to incentives and reward systems?

Answering such questions will require new collaborations between biologists and computer scientists, sociologists and mathematicians, physicists and robotics engineers. As in the past, these collaborations may turn out to be tricky, as each team member discovers the lingo and culture of their partners in discovery. Still, as Julia Parrish put it, there's never been a better time to get involved and get things done. "I feel like we're right at the point where all the girls and boys at a junior high school dance have been standing along the sidelines examining each other," she said of her fellow researchers. "Now we're ready to dance."

Understanding Collective Intelligence*

Satnam Alag

Web Applications Are Undergoing a Revolution

In this post dot com era the web is transforming. Newer web applications trust their users, invite them to interact, connect them with others, gain early feedback from them and then use the collected information to constantly improve the application. Web applications that take this approach develop deeper relationships with their users, provide more value to users as they return more often, and ultimately offer more targeted experiences for each user according to his or her personal need.

Web Users Are Undergoing a Transformation

Users are expressing themselves. This expression may be in the form of sharing their opinion on a product or a service through reviews or comments; through sharing and tagging content; through participation in an online community; or by contributing new content.

This increased user-interaction and participation gives rise to data that can be converted into intelligence in your application. The use of collective intelligence of users to personalize a site for a user, to aid him in search and to make decisions, to make the application more sticky are much cherished goals that web applications try to fulfill.

In his book, *Wisdom of the Crowds*, James Surowiecki, business columnist for The New Yorker, asserts that "under the right circumstances, groups are remarkably intelligent, and are often smarter than the smartest people in them." Surowiecki says that if the process is sound, the more people you involve in solving a problem, better will be the result. A crowd's "collective intelligence" will produce better results than those of a small group of experts if four basic conditions are met. These four basic

---

* The contents of this article are from Chapter 1 of the book *Collective Intelligence in Action* by Satnam Alag, published in October 2008 by Manning
  (http://www.manning.com/alag/).

J. Altmann et al. (Eds.): Advances in Collective Intelligence 2011, AISC 113, pp. 5–22.

conditions are: "Wise crowds" are effective when they are composed of individuals
who have diverse opinions; when the individuals are not afraid to express them;
there is diversity in the crowd and there is a way to aggregate all the information
and use it in the decision making process.

Collective intelligence is about making your application more valuable by tapping into "Wise crowds". More formally, Collective Intelligence (CI) as used in this
book simply and concisely means,

"To effectively use the information provided by others to improve one's application".

This is a fairly broad definition of collective intelligence – one which makes use
of all types of information, both inside and outside the application – to improve
the application for a user. This book introduces you to concepts from the areas of
machine learning, information retrieval, and data mining and demonstrates how you
can add intelligence to your application. You will be exposed to how your application can learn about individual users by using their interactions and correlate their
interactions with those of others to offer a highly personalized experience.

This chapter provides you with an overview of collective intelligence and how it
can manifest itself in your application. This chapter begins with a brief introduction
to the field of collective intelligence; then goes on to describe the many ways it can
be applied to your application; and finally how intelligence can be classified.

1 What Is Collective Intelligence (CI)?

Collective Intelligence is an active field of research that predates the web. Scientists from the fields of sociology, mass-behavior, and computer science have made
important contributions to this field. When a group of individuals collaborate or
compete with each other, intelligence or behavior that otherwise did not exist sud-
denly emerges, this is commonly known as Collective Intelligence. The actions or
influence of a few individuals slowly spreads across the community until it becomes
the norm. To better understand how this circle of influence spreads let us look at a
couple of examples.

In his book the 100th Monkey[1], Ken Keyes recounts an interesting story about
how change is propagated in groups. In 1952, on the isolated Japanese island of
Koshima scientist observed a group of monkeys. They offered them sweet potatoes,
which the monkeys liked but found the taste of dirt and sand on the potatoes un-
pleasant. One day an eighteen-month-old monkey, named Imo, found a solution to
the problem by washing the potato in a nearby stream of water. She taught her trick
to her mother. Her playmates also learned the trick and taught it to their mothers. Initially, only adults who imitated their children learned the new trick, while the others
continued eating the old way. In the autumn of 1958, there were a number of monkeys who were washing their potatoes before eating. The exact number is unknown,
but let's just say that out of a 1000 there were 99 monkeys who were washing their
potatoes before eating. Early one sunny morning a 100th monkey decided to wash

[1] http://en.wikipedia.org/wiki/Hundredth_Monkey

his potato. Then incredibly by evening *all* monkeys were washing their potatoes. The 100th monkey is that *tipping point* that caused others to change their habits for the better. Soon, it was observed that monkeys on other islands were also washing their potatoes before eating them.

As users interact on the web and express their opinions they influence others. Their initial circle of influence is the group of individuals that they most interact with. Because the web is a highly connected network of sites, This circle of influence grows and may shape the thoughts of everybody in the group. This *circle of influence* also grows rapidly throughout the community – another example helps illustrate this.

In 1918, as the influenza flu pandemic spread nearly 14% of Fiji's population died in just 16 days; nearly one third of the native population in Alaska had a similar fate; it is estimated that world-wide nearly twenty five million people died of the flu. A pandemic is a global disease outbreak and spreads from person to person. First, one person is affected, who then transmits it to another and then another. The newly infected person transmits the flu to others; this causes the disease to spread exponentially.

In October 2006, Google bought YouTube for $1.65 billion. In its 20 months of existence YouTube had grown to be one of the busiest sites on the internet, dishing out 100 million video[2] views a day. It ramped from zero to more than 20 million unique user visits a day, with mainly *"viral marketing"* – spread from person to person, similar to the way the pandemic flu spreads. In their case, each time a user uploaded a new video, he was easily able to invite others to view this video. As those others viewed this video, other related videos popped up as recommendations, keeping him further engaged. Ultimately, many of these "viewers" also became "submitters" and uploaded their own videos as well. As the number of videos increased the site became more and more attractive for new users to visit.

Whether you are a budding startup or, a recognized market leader, or an emerging application looking to take your application or web site to the next level, harnessing information from users improves the perceived value of the application to both current and prospective users. This improved value will not only encourage current users to interact more, but will also attract new users to the application. The value of the application further improves as new users interact with it and contribute more content. This forms a self-reinforcing virtuous feedback loop, commonly known as *Network Effect*, which enables wider adoption of the service. Next, let us look at CI as it applies to web applications.

2   CI in Web Applications

In this section we will look at how CI manifests itself in web applications. We will walk through an example to illustrate how it can be used in web applications, briefly review its benefits, see how it fits in with Web 2.0 and can be leveraged to build user-centric applications.

---

[2] As of September 2006.

Let's expand on our earlier definition of collective intelligence.

- the *intelligence* that is extracted out from the *collective* set of interactions and contributions made by your users.
- the use of this intelligence to act as a *filter* for what is *valuable* in your application for a user. This *filter* takes into account a user's preferences and interactions to provide relevant information to the user.

This filter could be the simple influence that collective user information has on a user - perhaps a rating or a review written about a product as shown in Figure 1 or it maybe more involved – building models to recommend personalized content to a user.

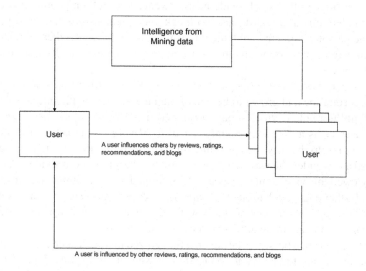

Fig. 1 A user may be influenced by other users either directly or through intelligence derived by the application by mining the data

As shown in Figure 2, there are three things that need to happen to apply collective intelligence in your application. You need to

1. allow users to interact with your site and with each other, learning about each user through their interactions and contributions
2. aggregate what you learn about your users and their contributions using some useful models
3. leverage those models to recommend relevant content to a user

Let us walk through an example to understand how collective intelligence can be a catalyst to building a successful web application.

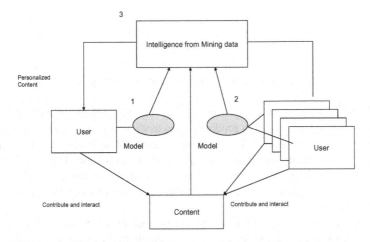

Fig. 2 Three components to harnessing collective intelligence: a. allow users to interact b. learn about your users in aggregate c. personalize content for a user using user interaction data and aggregate data

2.1 Collective Intelligence from the Ground Up: A Sample Application

In our hypothetical example, John and Jane are two engineers who gave up their lucrative jobs to start a company. They are based in the Silicon Valley and as is the trend nowadays they are building their fledgling company without any venture capital on a shoe-string budget leveraging open-source software. They believe in fast-iterative agile-based development cycles and are not afraid to release beta[3] software to gain early feedback on their features. They are looking to build a marketplace and plan to generate revenue via both advertisement and revenue share from selling items.

In their first iteration, they launched the application whereby users – mainly friends and family – could buy items and view relevant articles. There wasn't much in terms of personalization or user interaction or intelligence – a plain vanilla system.

Next, they added the feature of showing a *List* of top items purchased by users, along with a List of recently purchased items. This is perhaps the simplest form of applying collective intelligence – *providing information in the aggregate to users.* To grow the application virally, they also enabled users to email these *Lists* to others. *Users* used this to *forward* interesting list of items to their friends, who in turn became users of the application.

In their next iteration, they wanted to learn more about the user. So they built a basic user profile mechanism that contained explicit and implicit profile information. The explicit information was provided directly by the users as part of their account – first name, age, etc. The implicit information was collected from the *user*

---

[3] Note beta doesn't mean poor quality; it just means that it is incomplete in functionality.

interaction data – this included information such as the articles and content they viewed, and the products they purchased. They also wanted to show more relevant articles and content to each user, so they built a content-based *recommendation engine* that analyzed the content of articles – keywords, word frequency, location, etc. to *correlate* articles with each other and recommend possibly interesting articles to each user.

Next, they allowed users to generate content. They gave them an ability to write about their experiences with the products, in essence write *reviews* and create their list of *recommendations* through both explicit ratings of individual products and a "my top 10 favorite product" list. They also gave them a capability to *rate* items and *rate reviews*. Ratings and reviews have been shown to influence other users and numerical rating information is also useful as an input to a collaborative-based recommendation engine.

With the growing list of content and products available on the site, John and Jane now found it too cumbersome and expensive to manually maintain the classification of content on their site. The users also provided feedback that content navigation menus were too rigid. So they introduced dynamic navigation via a *Tag Cloud* – navigation built by an alphabetical listing of terms, where the font size correlates with importance or number of occurrences of a tag. The terms were automatically extracted from the content by analyzing the content. The application analyzed each user's interaction and provided the users with a personalized set of tags for navigating the site. The set of tags changed as the type of content visited by the users changed. Further, the content displayed when a user clicked on a tag varied from user to user and changed over time. Some tags pulled the data from a search engine, while others from the recommendation engine and external catalogs.

In the next release, they allowed the users to explicitly *tag* items by adding free text labels – along with saving or *book marking* items of interest. As users started *tagging* items, John and Jane found that there was a rich set of information that could be derived. First of all, users were providing new terms for the content that made sense to them – in essence they built *"folksonomies*[4]*"*. The tag cloud navigation now had both machine-generated and user-generated tags. The process of extracting tags using an automated algorithm could also be enhanced using the dictionary of tags built by the users. These user-added tags were also useful for finding keywords used by an ad-generation engine. They could also use the tags created by users to connect users with other user and with other items of interest. This is collective intelligence in action.

Next, they allowed their users to generate content. Users could now *blog* about their experiences, or ask and respond to *questions* on *message boards*, or participate in building the application itself by contributing to *wikis*. John and Jane quickly built an algorithm that could *extract tags from the unstructured content*. They then matched the interests of users – gained from analyzing their interaction in the applications – with *those of other users* to find *relevant* items. They were soon able to learn enough about their users to personalize the site for each user, to provide

---

[4] Folksonomies are classifications created through the process of users tagging items.

Table 1 Some of the ways to harness collective intelligence in your application

Techniques	Description
Aggregate Information: Lists	Create Lists of items generated in the aggregate by your users. Perhaps, Top List of items bought, or Top Search Items or List of Recent Items
Ratings, Review and Recommendations	Collective information from your users influences others
User generated content – blogs, wikis, message boards	Intelligence can be extracted from contributions by users. This contribution also influences other users.
Tagging, book marking, voting, saving	Collective intelligence of users can be used to bubble up interesting content, learn about your users and connect users.
Tag Cloud navigation	Dynamic classification of content using terms generated via one or more of the following techniques: machine-generated, professionally-generated, or user-generated
Analyze content to build user-profiles	Analyze content associated with a user to extract keywords. This information is used to build user-profile
Clustering and Predictive Models	Cluster users and items, build predictive models
Recommendation engines	Recommend related content or users based on intelligence gathered from user interaction and analyzing content
Search	Show more pertinent search results using a user's profile
Harness external content	Provide relevant information from the Blogosphere and external sites

relevant content – *targeting niche items to niche users*. They could also target relevant advertisements based on the user profile and context of interaction.

They also modified the *search* results to make it more *relevant* to a user for which they used the profile and interaction history of the user when appropriate. They also customized *advertising* by using keywords that were relevant to both the user and the page content.

To make the application stickier, they started aggregating and indexing external content – they would *crawl* a select list of external web sites to index the content and present links to it when relevant. They also connected to sites that tracked the blogosphere, presenting the users with relevant content from what others were saying in blogs.

They also clustered users and items to find patterns in the data and built models to automatically classify content into one of a many categories.

The users soon liked the application so much that they started recommending the application to their friends and relatives and the user base grew *virally*.

In our fictitious example, after a couple of years John and Jane retired to Hawaii, having sold the company for a gigantic amount, where they waited for the next web revolution ... Web 3.0!

This in essence are the many ways by which collective intelligence will manifest itself in your application and is more or less the outline for this book. Table 1 summarizes the ways to harness collective intelligence in your application.

John and Jane showed us a few nice things to apply to their site but there are other benefits of applying collective intelligence to your application. Let us look at that next.

2.2 Benefits of Collective Intelligence

Applying collective intelligence to your application impacts it in the following manner:

- Higher retention rates: The more users interact with the application – the stickier it gets for them – the higher is the probability that users will become repeat visitors.
- Greater opportunities to market to the user: The greater the number of interactions, the greater the number of pages visited by the user, which increases the opportunities to market to or communicate with the user.
- Higher probability of a user completing a transaction and finding information of interest: The more contextually relevant information that a user finds better are the chances that he will have the information he needs to complete the transaction or find content of interest. This will lead to higher click-through and conversion rates for your advertisements.
- Boosting search engine rankings: The more the user participate and contribute content the more content is available in your application and indexed by search engines. This could boost your search engine ranking and make it easier for others to find your application.

Collective intelligence is a term that is exceedingly being used in the context of Web 2.0 applications. Let's take a closer look at how it fits in with Web 2.0.

2.3 CI Is the Core Component of Web 2.0

"Web 2.0" is a term that has generated passionate emotions ranging from being dismissed as a marketing jargon to being anointed as the new or next generation of the internet. Using examples and commonality among new and older transformed companies, there are seven principles that Web 2.0 companies demonstrate as shown in Table 2[5].

It is widely regarded that "harnessing collective intelligence" is the key or core component to Web 2.0 applications. In essence, Web 2.0 is all about inviting users to participate and interact. But what do you do with all the data collected from user participation and interaction? This information is wasted if it cannot be converted into intelligence and channeled into improving one's application. This is where collective intelligence and this book come in.

Dion Hinchliffe in his article, *Five Great Ways to Harness Collective Intelligence*, makes an analogy to Einstein's belief that compound interest was the most important

---

[5] Refer to Tim O'Reilly's paper on Web 2.0.

Table 2 Seven Principles of Web 2.0 Applications

Principle	Description
The Network is the Platform	Companies or users who use traditional licensed software have to deal with running the software, upgrading it periodically to keep up with newer versions, and scaling it to meet appropriate levels of demand. Most successful Web 2.0 companies no longer sell licensed software, but instead deliver their software as a service. The end customer simply uses the service through a browser. All the headaches of running, maintaining and scaling the software and hardware are taken care of by the service provider seamless to the end user. The software is upgraded fairly frequently by the service provider and is available 24 x 7.
Harnessing Collective Intelligence	The key to the success of Web 2.0 applications is how effectively they can harness the information provided by the user. The more personalized your service, the better you can match a user to content of his choice.
Hard-to-Replicate Data as Competitive Advantage	Hard to replicate, unique, large datasets provides a competitive advantage to a company. Web 2.0 is a combination of "data" and "software". One cannot replicate Craigslist, eBay, Amazon, Flickr, or Google simply by replicating the software. The underlying data that the software generates from user activity tremendously valuable. This dataset grows everyday, improving the product daily.
The Perpetual Beta	Web 2.0 companies release their products early to involve their users and gain important feedback. They iterate often by having short release cycles. They involve the users early in the process. They instrument the application to capture important metrics on how a new feature is being used, how often it is being used, and by whom. If you are not sure on how a particular feature should look and have competing designs, expose a prototype of each to different sets of users and measure success of each. Involve the customers and let them decide which one they like. By having short development cycles it is possible to solicit user feedback, incorporate changes early in the product life cycle and build what the users really want.
Simpler Programming Models	Simpler development models lead to wider adoption and reuse. Design your application for "hackability" and "remixability" following open standards, using simple programming models and a licensing structure that puts as few restrictions as necessary.
Software above the Level of a Single Device	Applications that operate across multiple devices will be more valuable than those that operate in a single device.
Rich User Experience	The success of AJAX has fueled the growing use of rich user interfaces in Web 2.0 applications. Adobe Flash/Flex and Microsoft Silverlight are other alternatives for creating rich UIs.

force in the universe. Similarly web applications that effectively harness collective intelligence can "benefit" in much the same way - "harnessing collective intelligence is about those very same exponential effects".

COLLECTIVE INTELLIGENCE IS THE HEART OF WEB 2.0 APPLICATIONS.

It is also generally acknowledged that one of the core components of Web 3.0 applications will be the use of artificial intelligence[6]. There is debate as to whether this intelligence will be attained by computers reasoning like humans or by sites leveraging the collective intelligence of humans using techniques such as collaborative filtering. Either way, having the dataset generated from real human interactions will be necessary and useful.

In order to effectively leverage collective intelligence you need to put the user at the center of your application, in essence build a user-centric application.

2.4 Harnessing CI to Transform from Content-Centric to User-Centric Applications

Prior to the user-centric-revolution, many applications put very little emphasis on the user. These applications, known as content-centric applications, focused on the best way to present the content and were generally very static from user-to-user and from day-to-day. User-centric applications leverage CI to fundamentally change how the user interacts with the web application. User-centric applications make the user the center of the web experience and dynamically reshuffle the content based on what is known about the user and what the user explicitly asks for.

As shown in Figure 3, user-centric applications are composed of the following four components

- Core competency – the main reason why a user comes to the application.
- Community – connecting users with others users of interest, social networking, finding other users who may provide answers to a user's questions.
- Leveraging user-generated content – incorporating generated content and interactions of users to provide additional content to users
- Building a marketplace – monetize the application by product and/or service placements and showing relevant advertisements.

The user profile is at the center of the application. A part of the user profile may be generated by the user, while some parts of it may be learnt by the application based on the user interaction. Typically, sites that allow user-generated content have an abundance of information. User-centric sites leverage collective intelligence to present relevant content to the user.

---

[6] cf.http://en.wikipedia.org/wiki/Web_3.0#An_evolutionary_path_to_artificial_intelligence

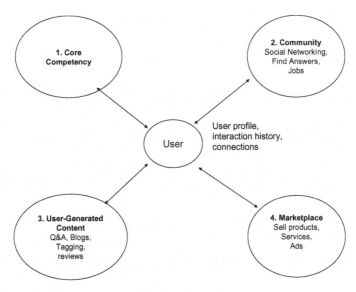

Fig. 3 Four pillars for user-centric applications

Figure 4 shows a screenshot of one such user-centric application – LinkedIn[7], a popular online network of more than 100 million professionals[8]. As shown in the screenshot, the LinkedIn application leverages the four components of user-centric applications

- Core competency: users come to the site to connect with others and build their professional profiles.
- Community: users create connections with other users; connections are used while looking up people, responding to jobs and answer questions asked by other users. Other users are automatically recommended as possible connections by the application.
- User-generated content: most of the content at the site is user-generated. This includes the actual professional profiles, the questions asked, the feed of actions – such as a user updating his profile, uploading his photograph, or connecting to someone new.
- Marketplace: the application is monetized by means of advertisements, job postings, and a monthly subscription by the power-users of the system, who often are recruiters. The monetization model used is also commonly known as "freemium[9]" – basic services are free and are used by most users, while there is a charge for premium services that a small minority of users pays for.

---

[7] http://www.linkedin.com/static?key=company_info&trk=ftr_abt
[8] As of July 2011.
[9] http://en.wikipedia.org/wiki/Freemium_business_model

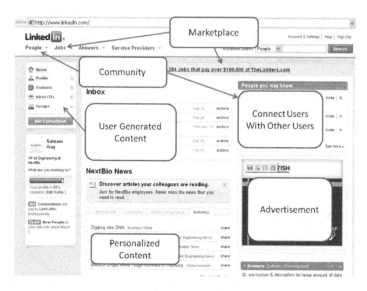

Fig. 4 An example of a user-centric application - LinkedIn (www.linkedin.com)

For user-centric applications to be successful they need to personalize the site for each user. CI can be very beneficial to these applications. So far in this section, we have looked at what is collective intelligence, how it manifests itself in your application, the advantages of applying it and how it fits in with Web 2.0. Next, we will take a more detailed look at the many forms of information provided by the users.

3   Classifying Intelligence

Figure 5 illustrates the three types of intelligence that we will discuss in this book. First, is explicit information that the user provides in the application. Second, is implicit information that a user provides either inside or outside the application and is typically in an unstructured format. Lastly, there is intelligence that is derived by analyzing the aggregate data collected. This piece of derived intelligence is on the upper half of the triangle, as it is based on the information gathered by the other two parts.

Data comes in two forms: structured data and unstructured data. Structured data has a well defined form, something that makes it easily stored and queried on. User ratings, content articles viewed and items purchased are all examples of structured data. Unstructured data is typically in the form of raw text. Reviews, discussion forum posts, blog entries, and chat sessions are all example of unstructured data.

In this section, we will look at the three forms of intelligence: explicit, implicit, and derived.

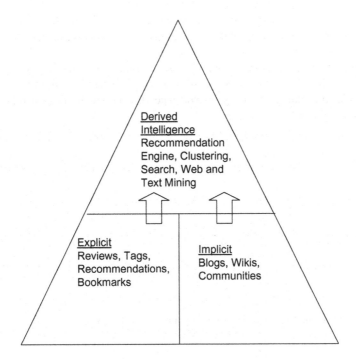

Fig. 5 Classifying user generated information

3.1 Explicit Intelligence

This section deals with explicit information that a user provides. Here are a few examples of how a user provides explicit information that can be leveraged.

Reviews and Recommendations

A recommendation made by a friend or a person of influence for can have a big impact on other users within the same group. Moreover, a review or comments about an experience of a user with a particular provider or service is contextually very relevant for other users inquiring about that topic, especially if it is within the context of similar use.

Tagging

Adding the ability for users to add tags - keywords or labels provided by a user – to classify items of interest such as, articles, items being sold, pictures, videos, podcasts, etc is a powerful technique to solicit information from the user. Tags can also be generated by professional editors or by an automated algorithm by analyzing the contents. These tags are used to classify data, bookmark sites, connect people with other people, to aid users in their search, and build dynamic navigation in your application of which a tag cloud is one example.

Figure 6 shows a tag cloud showing popular tags at del.icio.us, a popular bookmarking site. In a tag cloud, tags are displayed alphabetically with the size of the font representing the frequency of occurrence. The larger the font of the tag the higher is its frequency of occurrence.

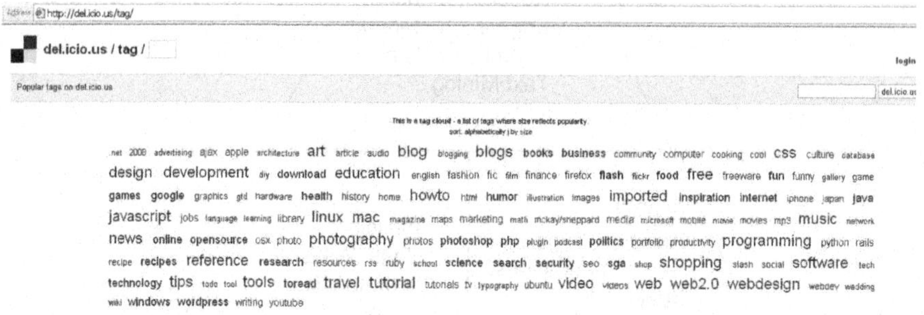

Fig. 6 This tag cloud from del.icio.us shows popular tags at the site

Voting

Voting is another way to involve and obtain useful information from the user. Digg, a website that allows user to contribute and vote on interesting articles, leverages this idea. Every article on Digg is submitted and voted on by the Digg community. Submissions that receive many votes in a short time period tend to move up in rank. This is a good way to share, discover, bookmark, and promote important news. Figure 7 is a screenshot from Digg.com showing news items with the number of Diggs associated with each.

3.2 Implicit Intelligence

This section deals with indirect information that a user provides. Here are a few examples of how a user provides this information.

Information relevant to your application may appear in an unstructured free-form text-format through reviews, messages, blogs, etc. A user may express his opinion online, either within your application or outside the application, by writing in his blog or replying to a question in an online community. Thanks to the power of search engines and blog tracking engines this information becomes easily available to others and helps to shape their opinions.

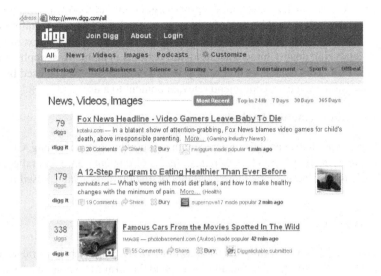

Fig. 7 Screenshot from Digg.com showing news items with the number of diggs for each

You may want to augment your current application by aggregating and mining external data. For example, if you area is real estate application you may want to augment your application with additional data harvested from freely available external sites, for example public records on housing sales, reviews of schools and neighborhoods, etc.

Blogs are online journals where information is displayed in reverse chronological order. The Blogosphere – the collection of blogs on the net – is huge and growing fast. As of March 2008, Technorati, a private company that tracks blogs, is tracking 112.8 million blogs. With a new blog being created virtually every second, the blogosphere is an important source of information that can be leveraged in your application. People write blogs on virtually every topic.

Next, let us look at the third category of intelligence, which is derived from analyzing the data collected.

3.3 Derived Intelligence

This section deals with information derived from the data collected. Here are a few examples of techniques and features that deal with derived intelligence.

Data and Text Mining

The process of finding patterns and trends, which would otherwise go undetected, in large datasets using automated algorithms is known as data mining. When the data is in the form of text, the mining process is commonly known as text data mining. Another related field is information retrieval, which deals with finding relevant information by analyzing the content of the documents. Web and text mining deal with analyzing unstructured content to find patterns in them. Most applications are content rich. This content is indexed by search engines and can be used by the recommendation engine to recommend relevant content to a user.

Clustering and Predictive Analysis

Clustering and predictive analysis are two main components of data mining. Clustering techniques enable you to classify items – users or content – into natural groupings. Predictive analysis is a mathematical model that predicts a value based on the input data.

Intelligent Search

Search is one of the most commonly used techniques used by users to retrieve content. Applying an ontology to automatically expand the term for synonyms, leveraging citation or linking information to the content, leveraging parts of speech tagging and NLP are all examples of intelligent search

Recommendation Engine

A recommendation engine offers *relevant* content to a user. Again, recommendation engines can be built by either analyzing the content, by analyzing user-interactions (collaborative approach) or a combination of both. Figure 8 shows a screenshot from Yahoo! Music in which a user is recommended music by the application.

Fig. 8 Screenshot from Yahoo! Music recommending songs of interest

Recommendation engines use inputs from the user to offer them a list of recommended items. The inputs to the recommendation engine maybe items in the user's shopping list, items they have purchased in the past or are considering purchasing, user profile information such as age, tags and articles that the user has looked at or contributed, or any other useful information that the user may have provided. For large online stores such as Amazon, which has millions of items in its catalog, providing fast recommendations can be challenging. Recommendation engines need to be fast and scale independent of the number of items in the catalog and the number of users in the system; they need to offer good recommendations for new customers with limited interaction history; and they need to age out older or irrelevant interaction data (such as a gift bought for someone else) from the recommendation process.

4 Summary

Collective intelligence is powering a new breed of applications that invite the users to interact, contribute content, connect them with other users, and personalize the site experience.

Users influence other users. This influence spreads outwards from their immediate circle of influence until it reaches a critical number, after which is becomes the norm. Useful user-generated content and opinions spread virally with very minimal marketing.

Intelligence provided by the users can be divided into three main categories. The first is direct information/intelligence provided by the user. Reviews, recommendations, ratings, voting, tags, bookmarks, user interaction, and user generated content are all examples of techniques to gather this intelligence. Next follows indirect information provided by the user either on or off the application, which is typically in unstructured text. Blog entries, contributions to online communities, wikis, are all sources of intelligence for the application. The third is a higher level of intelligence that is derived using data mining techniques. Recommendation engines, use of predictive analysis for personalization, profile building, market segmentation, web and text mining are all examples of discovering and applying this higher level of intelligence.

References

1. Alag, S.: Collective Intelligence in Action. Manning, Greenwich (2008), http://www.manning.com/alag/
2. All things Web 2.0, http://www.allthingsweb2.com/component/option,com_mtree/Itemid,26/
3. Anderson, C.: The Long Tail: Why the Future of Business Is Selling Less of More. Hyperion, New York (2006)
4. BBC news: One blog created every second, http://news.bbc.co.uk/1/hi/technology/4737671.stm

5. Hinchcliffe, D.: The Web 2.0 Is Here,
   `http://web2.wsj2.com/web2ishere.htm`

6. Hinchcliffe, D.: Architectures of Participation: The Next Big Thing (2006),
   `http://web2.wsj2.com/architectures_of_participation_the_next_big_thing.htm`

7. Hinchcliffe, D.: Five Great Ways to Harness Collective Intelligence (2006),
   `http://web2.wsj2.com/five_great_ways_to_harness_collective_intelligence.htm`

8. Horrigan, J., Rainie, L.: The Internet's Growing Role in Life's Major Moments (2005),
   `http://www.pewinternet.org/pdfs/PIP_Major%20Moments_2005.pdf`

9. Jaokar, A.: Tim O' Reilly's seven principles of web 2.0 make a lot more sense if you change the order (2006),
   `http://opengardensblog.futuretext.com/archives/2006/04/tim_o_reillys_s.html`

10. Kroski, E.: The Hype and Hullabaloo of Web 2.0,
    `http://infotangle.blogsome.com/2006/01/13/the-hype-and-the-hullabaloo-of-web-20/`

11. McGovern, G.: Collective intelligence: is your website tapping it? In: New Thinking (2006),
    `http://www.gerrymcgovern.com/nt/2006/nt-2006-04-17-collective-intelligence.htm`

12. Online Community Toolkit,
    `http://www.fullcirc.com/community/communitymanual.htm`

13. O'Reilly, T.: The Future of Technology and Proprietary Software (2003),
    `http://tim.oreilly.com/articles/future_2003.html`

14. O'Reilly, T.: What Is Web 2.0: Design Patterns and Business Models for the Next Generation of Software (2005),
    `http://www.oreilly.com/pub/a/oreilly/tim/news/2005/09/30/what-is-web-20.html`

15. O'Reilly, T.: Web 2.0: Compact Definition? (2005),
    `http://radar.oreilly.com/archives/2005/10/web_20_compact_definition.html`

16. Por, G.: The meaning and accelerating the emergence of CI (2004),
    `http://www.community-intelligence.com/blogs/public/archives/000251.html`

17. Surowiecki, J.: The Wisdom of Crowds: Why the Many are Smarter than the Few and How Collective Wisdom Shapes Business, Economics, Societies and Nations. Anchor, New York (2005)

18. Wikipedia: Web 3.0,
    `http://en.wikipedia.org/wiki/Web_3.0#An_evolutionary_path_to_artificial_intelligence`

Predicting Asset Value through Twitter Buzz

Xue Zhang, Hauke Fuehres, and Peter A. Gloor

Abstract. This paper describes early work trying to predict financial market movement such as gold price, crude oil price, currency exchange rates and stock market indicators by analyzing Twitter posts. We collected Twitter feeds for 5 months obtaining a large set of emotional retweets originating from within the US, from which six public opinion time series containing the keywords "$dollar\%_t$", "$\$\%_t$", "$gold\%_t$", "$oil\%_t$, "$job\%_t$" and "$economy\%_t$" were extracted. Our results show that these variables are correlated to and even predictive of the financial market movement. Except "$\$\%_t$", all other five public opinion time series are identified by a Granger-causal relationship with certain market movements. It is demonstrated that daily changes in the volume of economic topic retweeting seem to match the value shift occurring in the corresponding market next day.

1 Introduction

"Prediction is difficult, especially when its about the future" – Niels Bohr

We human beings are always curious about the future. For thousands of years, people have been trying their best to predict what would happen next. Although most of these predictions turn out to be wrong, people never give up predicting, and are continuously trying to improve it. Weather forecasts, earthquake early warnings, stock market predictions etc. – predictions are an important part in everyday life. With the popularization of Internet and online social networking, this time-honored activity enters a new era.

Xue Zhang
Department of Mathematic and Systems Science, National University of Defense Technology, Changsha, Hunan, P.R.China

Xue Zhang · Hauke Fuehres · Peter A. Gloor
MIT Center for Collective Intelligence, Cambridge MA, USA

J. Altmann et al. (Eds.): Advances in Collective Intelligence 2011, AISC 113, pp. 23–34.
springerlink.com   © Springer-Verlag Berlin Heidelberg 2012

Recently, a lot of research has been done on prediction with data from social networks and web searches. Gayo-Avello et al. [7] clearly pointed out that following what people are blogging about or what they are searching about can give us some intuition on the collective psyche and lead us to understand what is currently happening in society before it is actually happening. Sometimes people refer to this phenomenon as the "wisdom of the crowd", that is, taking into account the opinion of the society as a whole, instead of the opinion of the expert.

A group of researchers is applying this novel methodology to stock market prediction. Antweiler and Frank [1] determined correlation between activity in Internet message boards and stock volatility. Gilbert and Karahalios [8] used over 20 million posts from the LiveJournal website to create an index of the US national mood, which they call the Anxiety Index. They found that when this index rose sharply, the S&P 500 ended the day marginally lower than is expected. Choudhury et al. [5] modeled contextual properties of posts in SVMs (support vector machines) and trained it with stock movement. The result shows about 87% accuracy in predicting the direction of the movement.

As one of the most popular social networking websites, Twitter is drawing more and more attention from researchers from different disciplines. There are several streams of research investigating the role of Twitter. One stream of research focuses on understanding its usage and community structure [4, 9, 10, 18]. Other researchers are more interested in its prediction power and potential application in other areas. It has been demonstrated that by tracking tweet numbers related to certain topics, both box-office revenues of movies and political elections could be successfully forecasted [2, 17]. Also, Twitter has been used in tracking the spread of epidemic disease [11].

Twitter buzz was also employed in predicting the stock market movement. By analyzing the sentiment of a random sample of tweets, Bollen et al. [3] found that public mood can be used to predict the stock market. Furthermore, stock-related tweets with a specific hashtag "$" were collected and studied in detail in [16], where it was found that these tweets contain valuable information that is not fully incorporated in current market indicators. In previous work [19], we also presented very preliminary results that the number of emotional tweets, which contain words such as "hope", "fear" or "worry" correlated with stock market indicators. In this paper, further tests and analysis to predict valuation of tradable assets will be described.

The rest of the paper is organized as follows. In section 2, we present our Twitter dataset and financial market dataset, laying out how we constructed the public opinion time series. Section 3 discusses the method of determining correlation between Twitter buzz and market movement and presents the results, which are followed by discussion and future work in section 4.

2 Data

Twitter is a worldwide popular website, which offers a social networking and microblogging service, enabling its users to update their status in tweets, follow the people they are interested in, retweet others' posts and even communicate with them directly. Since it launched in 2006, its user base has been growing exponentially. As of June 2011, it is estimated to have 200 million users, generating 190 million tweets a day and handling over 1.6 billion search queries per day. The rising popularity of twitter gives us a novel way of capturing the collective mind up to the last minute. In this section, we introduce the datasets that form the basis of the work described in this paper.

2.1 Twitter Data Collection

We collected a large set of tweets submitted to Twitter in the period from November 15, 2010 to April 20, 2011. In order to get a better picture of the opinion and emotional state of the US investors, we only filter for emotional retweets that come from the United States. In other words, all the data we collected meets the following conditions:

Retweets Only

Structurally, retweeting is the Twitter-equivalent of email forwarding where users post messages originally posted by others. As an integral part of the Twitter experience, the retweeting phenomenon has been explicitly studied in prior research [4]. It is generally believed that the more a topic is being picked up and retweeted by others, the more it is relevant and widely recognized. Although there is no universally agreed-upon syntax for retweeting, "RT @user message" is the prototypical formulation where the referenced *user* is the original author and *message* is the original tweet's content, therefore we choose "RT @" as our indicator of retweets.

Containing the Emotion Words "Hope", "Fear" Or "Worry"

Emotional state greatly influences human decisions, which obviously include the appropriate choice of an investment strategy [6, 12, 13, 14, 15]. When people are pessimistic or uncertain about their future, they will be more cautious to invest and trade. Therefore capturing the collective mind – especially people's mood – becomes one possible way to predict the future. To be consistent with and further test previous work, we only take into account the retweets that contain the words "hope", "fear" or "worry", because we had found in earlier work [19] that these words are excellent indicators of emotion-laden tweets.

Originating from the US

The goal of this paper is to analyze whether Twitter buzz can be helpful in forecasting selected economic indicators of the US economy. For the purpose of better capturing the opinion and emotional state of the US population, we intentionally limit the targeted tweets to the ones originating from within the continental United States without Alaska. Tweets were collected within four 2000-kilometers circles with centers in Pittsburg, Atlanta, Las Vegas and Boise respectively. As Figure 1 shows, these circles cover the contiguous United States and parts of Canada and Mexico.

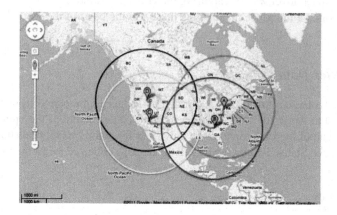

Fig. 1 Geographical origin of Twitter data: A is Pittsburg, B is Atlanta, C is Las Vegas and D is Boise

Over the duration of five months, 3,809,437 retweets posted by approximately 961,000 users were collected and each tweet has a unique identifier, time of submission and the textual content. Table 1 summarizes the daily number of retweets related to each emotional word. As we can see, the daily retweet rate of each emotion word is highly variable, for example, the hope-retweets range from 6453 to 34805 per day. Even more interestingly, the number of hope-retweets is much higher than the fear or worry ones, almost six times on average, which might suggest that people prefer using optimistic words when they express their feelings, even when they are worrying or in fear.

Table 1 Daily number of emotional retweets

	Average per day	Min per day	Max per day
Hope-retweet#	20613	6453	34805
Fear-retweet#	3710	853	7555
Worry-retweet#	3653	1071	7397
Total#	27977	11395	46209

2.2 Generating Public Opinion Time Series

In this section, we further discuss how to extract posts in regard to economic topics from our emotional retweets dataset. As twitter users can share only short textual messages with no more than 140 characters per post, there is always only one topic in one tweet. No matter if it is a piece of news announcing the death of Osama bin Laden or a conversation between two friends talking about the wonderful birthday party, owing to the length limitation, the tweet stays on the same topic. Thus the main theme of the whole tweet can usually be subsumed by one or two keywords.

Inspired by this property, a list of words related to economy was selected as a clue for economic tweets. This list of keywords includes *"dollar"*, *"$"*, *"gold"*, *"oil"*, *"job"* and *"economy"*. Then we measured collective opinion on each day by simply counting how many retweets contain these words. As the total number of retweets varies highly from day to day, a normalized number was chosen as a measurement of public opinion on day *t*. For example, we counted the number of retweets containing the word *"dollar"* and normalized it by the total retweet number on the same day *t*, this normalized retweet number is listed as *"dollar%$_t$"*. Figure 2 below illustrates all 6 public opinion time series.

2.3 Market Data

In this section, we look at different categories of assets including the gold price, crude oil price, currency exchange rates and stock market indicators. For our analysis we have taken the daily price of gold (dollars per ounce), WTI Cushing crude oil price (dollars per barrel), currency exchange rates (USD/CHF), Dow Jones Industrial Average (DJIA), NASDAQ and S&P 500 all collected during the same period as the Twitter data.

From Figure 3, we can see that although there is much fluctuation, the overall trend in this period of stock market, crude oil and gold price is all up. However, in contrast, the exchange rate of USD to CHF declined almost 10% at the same time. Obviously, all these market time series are non-stationary. To meet the requirement of stationarity in time series analysis, data are processed in the following way. Taking DJIA as an example, the stock movement at a day *t* is defined as the normalized change in stock close price from the past day, which can be expressed as

$$D_t = \frac{DJIA_t - DJIA_{t-1}}{DJIA_{t-1}} \tag{1}$$

where $DJIA_t$ is the close price of day *t*. Similarly, we determine the other independent variables. Using these new relative variables, we not only can tell the change direction of the market which is indicated by the sign of the number, but also measure how much it changed compared to the previous day.

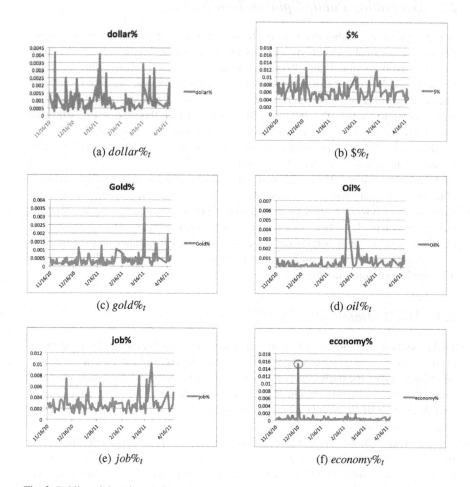

(a) $dollar\%_t$

(b) $\$\%_t$

(c) $gold\%_t$

(d) $oil\%_t$

(e) $job\%_t$

(f) $economy\%_t$

Fig. 2 Public opinion time series

$$N_t = \frac{NASDAQ_t - NASDAQ_{t-1}}{NASDAQ_{t-1}} \quad S_t = \frac{S\&P_t - S\&P_{t-1}}{S\&P_{t-1}}$$

$$O_t = \frac{Oil_t - Oil_{t-1}}{Oil_{t-1}} \quad G_t = \frac{Gold_t - Gold_{t-1}}{Gold_{t-1}} \qquad (2)$$

$$U_t = \frac{USD_t - USD_{t-1}}{USD_{t-1}}$$

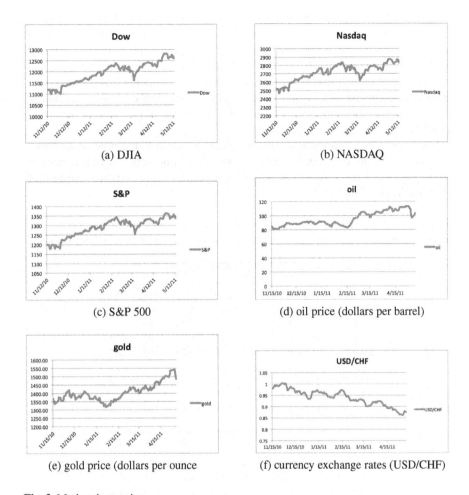

(a) DJIA   (b) NASDAQ

(c) S&P 500   (d) oil price (dollars per barrel)

(e) gold price (dollars per ounce   (f) currency exchange rates (USD/CHF)

Fig. 3 Market time series

3   Methods and Results

3.1   Correlation between Public Opinion and Market Time Series

To obtain a first indication whether the Twitter information might help forecast the asset value, we analyzed the correlation between the two time series. Tables 2 to 4 illustrate correlation coefficients between market movement on day t and Twitter buzz of day $t - i (i = 1, 2, 3)$ separately.

In Table 2, we observe a relatively strong correlation between stock market return and "$dollar\%_{t-1}$" ($r = 0.308^{**}, 0.203$ and $0.259^{*}, p$-value $= 0.004, 0.058$ and 0.015). In addition, not only "$oil\%_{t-1}$" but also "$economy\%_{t-1}$" is strongly

correlated with oil price changes of day t ($r = 0.295^{**}$ and 0.214^*, p-value $=$
0.006 and 0.046). Even more interestingly, we found that the correlation between
"$gold\%_{t-1}$" and G_t is weak, but "$gold\%_{t-1}$" is significantly correlated with U_t
($r = 0.213^*$, p-value $= 0.016$), indicating a relationship between the gold price and
the strength of the US dollar. Furthermore, it is worth noticing that all the correlation coefficients mentioned above are positive, which implies that an increase in
economic topic retweeting seems to indicate an increase in the value of the corresponding asset on the next market day.

In contrast, the relationships between market movement and time series "$\%$"
and "$job\%$" are not that significant in this period. Additionally, the Twitter buzz of
two or three days before seems to have less influence on the market movement of
day t (see Tables 3 and 4).

Table 2 Correlation coefficient between market movement and Twitter buzz 1 day before

	D_t	N_t	S_t	O_t	G_t	U_t
$dollar\%_{t-1}$	.308**	.203	.259*	.012	−.112	−.055
$\%_{t-1}$	.108	.062	.116	.004	−.080	−.122
$gold\%_{t-1}$	.122	.055	.088	−.034	−.053	.213*
$oil\%_{t-1}$	.022	.018	.054	.295**	.108	.072
$job\%_{t-1}$	−.035	.000	−.013	−.165	−.203	.167
$economy\%_{t-1}$	−.142	−.186	−.147	.214*	−.011	−.021

Table 3 Correlation coefficient between market movement and Twitter buzz 2 days before

	D_t	N_t	S_t	O_t	G_t	U_t
$dollar\%_{t-2}$	.106	.040	.065	−.148	−.078	.122
$\%_{t-2}$	.040	.025	.033	−.099	.077	−.131
$gold\%_{t-2}$	−.032	−.013	−.041	.020	.089	.064
$oil\%_{t-2}$	.004	−.039	−.019	.201	−.004	−.123
$job\%_{t-2}$	.094	.101	.108	−.151	−.116	.081
$economy\%_{t-2}$	−.073	−.068	−.030	.121	.020	−.039

Table 4 Correlation coefficient between market movement and Twitter buzz 3 days before

	D_t	N_t	S_t	O_t	G_t	U_t
$dollar\%_{t-3}$	.018	.010	.026	−.013	−.273*	−.088
$\%_{t-3}$	−.198	−.179	−.176	−.109	−.048	−.142
$gold\%_{t-3}$	.020	.024	.007	−.033	.030	.133
$oil\%_{t-3}$	.033	.077	.069	.039	−.039	−.017
$job\%_{t-3}$	.126	.156	.130	−.132	−.141	.093
$economy\%_{t-3}$	−.086	−.025	−.029	.031	.152	−.118

3.2 Granger-Causality Analysis

In this section we apply Granger causality analysis to the daily time series of public opinion vs. financial market movement. Granger causality is a statistical concept of causality that is based on prediction. According to Granger causality, if a signal X "Granger-causes" (or "G-causes") a signal Y, then past values of X should contain information that helps predict Y above and beyond the information contained in past values of Y alone. Its mathematical formulation is based on linear regression modeling of stochastic processes (Granger 1969). It is noteworthy that in spite of its name, Granger causality is not sufficient to imply true causality. If both X and Y are driven by a common third process with different lags, X might erroneously be believed to "Granger-cause" Y. However, in our project, we are not testing the actual causation but simply whether one variable provides predictive information about the other one or not.

Granger causality requires that the time series have to be covariance stationary, so an Augmented Dickey-Fuller test has been done first, in which the null hypothesis H_0 of non-stationarity was rejected at the 0.05 confidence level. Again, all Twitter buzz and market movement time series were verified to be stationary.

To test whether public opinion time series "Granger-cause" the changes in financial market valuation, two linear regression models were applied as shown in equations (3) and (4). The first model (M_1) uses only n lagged values of market data to predict Y_t, while the second model (M_2) also includes the lagged value of public opinion time series, which are denoted by $X_{t-1}, \ldots, X_{t-n}$. In order to find an appropriate number of lags, we set the lag parameter n equal to 1, 2 and 3 separately.

$$M_1 : Y_t = \alpha \sum_{i=1}^{n} \beta_i Y_{t-i} + \varepsilon_t \tag{3}$$

$$M_2 : Y_t = \alpha \sum_{i=1}^{n} \beta_i Y_{t-i} + \sum_{j=1}^{n} \gamma_j X_{t-j} + \varepsilon_t \tag{4}$$

After establishing the linear regression equations, a statistics f is defined as

$$f = \frac{\frac{SSR_1 - SSR_2}{n}}{\frac{SSR_2}{m - 2n - 1}} \tag{5}$$

where SSR_1 and SSR_2 are the two sum of squares residuals of equations (3) and (4); m is the number of observations. Theoretically, $f\ F(n, m - 2n - 1)$. Thus, the question whether X "Granger-causes" Y could be solved by simply checking the p-values.

From Table 5 we can easily draw the conclusion that Twitter buzz indeed has some information that can be used in predicting financial market movement. We observe that "$dollar\%_t$" has the highest Granger causality relation with stock market return, especially with the DJIA return (p-value < 0.01 when $n = 1$ and remains significant when $n = 2$ and 3). Also, the "$oil\%_t$" time series "Granger-causes" the changes in oil price (p-value is always less than 0.05 when lag varies from 1 to 3 days). The other two predictive variables are "$gold\%_t$" and "$job\%_t$", which have Granger causality relation with U_t and G_t separately. However, the most interesting aspect is that the "$gold\%_t$" time series failed in explaining the price change in the gold market, but could help predict the currency exchange rates (USD/CHF). We speculate that currency fluctuation, and the underlying lack in confidence in the national economy influence the eagerness of buyers to invest into the "safe haven" gold.

Table 5 Statistical significance (p-value) of bivariate Granger causality correlation between Twitter buzz and financial market movement (p-value < 0.05 :*, p-value < 0.01 :**)

	Lag	D_t	N_t	S_t	O_t	G_t	U_t
$dollar\%_t$	$n=1$	.0039**	.0494*	.0165*	.9588	.2877	.3131
	$n=2$	.0203*	.2405	.124	.3174	.3301	.1014
	$n=3$	.0272*	.3869	.1803	.4251	.0527	.0802
$\$\%_t$	$n=1$	.3309	.5538	.2966	.9033	.4887	.2315
	$n=2$	.9919	.9864	.9849	.7672	.6766	.357
	$n=3$	.2369	.3818	.3967	.8694	.7855	.3395
$gold\%_t$	$n=1$	.286	.5908	.4459	.7685	.5943	.0053*
	$n=2$	.3047	.7991	.572	.8669	.7787	.0306*
	$n=3$	.3267	.8883	.7072	.9317	.9964	.0518
$oil\%_t$	$n=1$	.9036	.7989	.6828	.0164*	.2359	.6347
	$n=2$	.8657	.4439	.7809	.0096**	.3672	.6156
	$n=3$	.9102	.5801	.864	.0225*	.3668	.3782
$job\%_t$	$n=1$	.7542	.9812	.9222	.1668	.0413*	.1167
	$n=2$	.2322	.3358	.204	.3986	.194	.3919
	$n=3$	.4879	.5408	.4544	.3697	.0896	.6271
$economy\%_t$	$n=1$	.1978	.0819	.183	.0558	.8752	.591
	$n=2$	.6916	.297	.5575	.1293	.9446	.7235
	$n=3$	.6592	.4267	.6904	.1896	.4743	.3417

4 Discussion

In this paper, we investigated the relationship between Twitter buzz and financial market movement. Our results statistically show that public opinion measured from large-scale collection of emotional retweets is correlated to and even predictive of the financial market movement. Except "$\$\%_t$", all other five public opinion time series are identified in a Granger-causal relationship with selected asset valuation movements. The changes in the volume of economic topic retweeting seems to match the value shift occurring in corresponding next market day.

However, there are still a number of important factors not acknowledged in our analysis to be studied in future work. First, unlike the prior work of [3] and [16], when we extracted the public opinion from Twitter, we neither constrained our data to those stock-related tweets which have a specific hashtag nor use sentiment analysis tools to measure the public mood from a random sample of tweets. We chose a few keywords to identify the emotional retweets talking about economic activity, then use volume change to track the public opinion. This method is simple and useful. It however does not linguistically analyze the content of tweets, which might offer additional valuable information. Advanced sentiment analysis could be employed in future work to improve our results. Second, the analyzing methods we used in this paper, both the correlation and Granger causality analysis, are based on the assumption that the relation between variables is linear, which is hardly satisfied for financial market movement. More advanced tools that can better characterize the non-linear relationship between variables, such as Neural Networks and Support Vector Machines, should also be explored in future work.

References

1. Antweiler, W., Frank, M.Z.: Is All That Talk Just Noise? The Information Content of Internet Stock Message Boards. Journal of Finance 59(3), 1259–1294 (2004)
2. Asur, S., Huberman, B.A.: Predicting the Future With Social Media. In: Proceedings of the 2010 IEEE/WIC/ACM International Conference on Web Intelligence and Intelligent Agent Technology (2010)
3. Bollen, J., Mao, H., Zheng, X.J.: Twitter mood predicts the stock market. Journal of Computational Science 2(1), 1–8 (2011)
4. Boyd, D., Golder, S., Lotan, G.: Tweet, Tweet, Retweet: Conversational Aspects of Retweeting on Twitter. In: Proceeding of 43rd Hawaii International Conference on System Sciences, HICSS (2010)
5. Choudhury, M.D., Sundaram, H., John, A., Seligmann, D.D.: Can Blog Communication Dynamics be Correlated with Stock Market Activity? In: Proceedings of the 9th ACM Conference on Hypertext and Hypermedia (2010)
6. Dolan, R.J.: Emotion, cognition, and behavior. Science 298(5596), 1191–1194 (2002)
7. Gayo-Avello, D., Metaxas, P.T., Mustafaraj, E.: On the Unpredictability of Elections using Social Media Data. In: Interdisciplinary Workshop on Information and Decision in Social Networks. MIT (2011)
8. Gilbert, E., Karahalios, K.: Widespread Worry and the Stock Market. In: Proceedings of the 4th International AAAI Conference on Weblogs and Social Media, ICWSM (2010)
9. Huberman, B.A., Romero, D.M., Wu, F.: Social networks that matter: Twitter under the microscope. First Monday 14(1) (2009)
10. Java, A., Song, X., Finin, T., Tseng, B.: Why We Twitter: Understanding Microblogging Usage and Communities. In: Proceeding of 9th WebKDD and 1st SNA-KDD Workshop on Web Mining and Social Network Analysis (2007)
11. Lampos, V., Cristianini, N.: Tracking the flu pandemic by monitoring the Social Web. In: IAPR 2nd Workshop on Cognitive Information Processing (2010)
12. Lerner, J.S., Keltner, D.: Fear, Anger, and Risk. Journal of Personality and Social Psychology 81(1), 146–159 (2001)

13. Lerner, J.S., Small, D.A., Loewenstein, G.F.: Heart strings and Purse Strings: Carryover Effects of Emotions on Economic Decisions. Psychological Science 15, 337–341 (2004)
14. Loewenstein, G.F., Weber, E.U., Hsee, C.K., Welch, N.: Risk as Feeling. Psychological Bulletin 127(2), 267–286 (2001)
15. Shiv, B., Loewenstein, G.F., Bechara, A., Damasio, H., Damasio, A.R.: Investment Behavior and the Negative Side of Emotion. Psychological Science 16, 435–439 (2005)
16. Sprenger, T.O., Welpe, I.M.: Tweets and Trades – The Information Content of Stock Microblogs (2010), SSRN: http://ssrn.com/abstract=1702854
17. Tumasjan, A., Sprenger, T.O., Sandner, P.G., Welpe, I.M.: Predicting Elections with Twitter: What 140 Characters Reveal about Political Sentiment. In: Proceedings of the 4th International AAAI Conference on Weblogs and Social Media, ICWSM (2010)
18. Wu, S., Hofman, J.M., Mason, W.A., Watts, D.J.: Who says What to Whom on Twitter. In: Proceedings of the 20th International Conference on World Wide Web (2011)
19. Zhang, X., Fuehres, H., Gloor, P.: Predicting Stock Market Indicators Through Twitter: I hope it is not as bad as I fear (2010),
    http://www.ickn.org/documents/COINs2010_Twitter4.pdf

Adding Value with Collective Intelligence – A Reference Framework for Business Models for User-Generated Content

Henrik Ickler and Ulrike Baumöl

Abstract. Many web-based business models started to harnessing collective intelligence by integrating users and customers into the value-adding process. With common web applications users have the chance to be a significant provider of contents and can commit themselves to deliver. Today, more and more companies detect the commercial potential of this user-generated content (UGC). Amazon, for example, is one of the pioneers in integrating customers and users for providing comments and evaluations. More and more, business models started to evolve which build on UGC at their core. These UGC business models define a certain business model type which bases its primary service offering on UGC.

This paper describes and analyzes building blocks of business models for UGC to be found in the existing literature and in case studies. For that, the research method of multiple case study analysis is employed. As a result, a reference framework for business models for UGC is proposed, which can be used for the definition, development, comparison as well as the assessment of such business models.

1 Introduction

O'Reilly [17] started the discussion of new ways how to use the Internet already in the year 2005 with his paper "What is Web 2.0". With that he introduced the term Web 2.0 and already hinted at the meaning of collective intelligence (CI). Harnessing CI promises a considerable potential for value adding and benefits in many different areas, e.g. economy, research, and society. There are already well-known examples for the potential benefit of using CI, such as Wikipedia. However, research on the basic mechanisms and a deeper understanding of CI is only in its beginnings [11] [15]. An important part of CI is the creation of content by customers as well

Henrik Ickler · Ulrike Baumöl
University of Hagen, Faculty of Economics, Department of Information Management,
Universitätsstr. 41, 58097 Hagen, Germany
e-mail: {henrik.ickler,ulrike.baumoel}@fernuni-hagen.de

J. Altmann et al. (Eds.): Advances in Collective Intelligence 2011, AISC 113, pp. 35–52.
springerlink.com   © Springer-Verlag Berlin Heidelberg 2012

as users, who are not yet customers, using the Internet as a platform and with that creating web-based CI. This is also called user-generated content (UGC) and is a defining characteristic of Web 2.0 and social commerce, which is a part of it [9] [10]. New and existing business models have started to integrate UGC to a certain degree into their production processes and sometimes even use it as primary service offering. Business models have been subject to research for some time now [20] [24], but business models for UGC are still not in the center of attention and are often only mentioned as a side note in connection with Web 2.0 business models [9]. However, if web-based CI is to be effectively harnessed for value-adding, it is mandatory to understand how business models have to be designed [13]. As a consequence, a first and important step is to understand already existing business models using web-based CI.

This paper contributes by providing a suggestion for closing this gap. It presents an analysis approach for variants of UGC business models, basing on a literature and multiple case study analysis. The result is a reference framework consisting of nine building blocks for respective business models and their characteristics as well as instances.

Based on the research question what the constituting characteristics of UGC business models are and how they can be integrated into a reference framework for this type of business models, the paper is organized as follows: After the introduction, a literature review is presented which aims at describing and defining relevant terms, such as UGC and business model. In the next part, the research method is briefly described before deducing and discussing the building block of the reference framework. Then, the framework is used to show how a concrete UGC business model can be described. The paper concludes with a brief summary.

2   Literature Review

2.1   User-Generated Content

Decreasing costs of use and more user-friendly applications led to a dramatic rise of Internet user, or rather Web users of the past few years. At the same time, a change of user behavior could be noticed [1]. The Web of today is characterized by participation and interaction. Users have changed from being passive consumers of content to active producers. Many users willingly provide content, such as comments and evaluations, pictures, or video clips, for other users. Terms used for this phenomenon are user-created content (UCC), consumer-generated media (CGM) or user-generated content (UGC) [32]. All these terms focus on the "consumer" or "user". A user is defined as a person who creates content and does that in an Internet-independent way. Moreover, he or she publishes it for addressees who are not fully specified beforehand [27]. This person is not a professional publisher and does not pursue commercial purposes [28]. Typical UGC are text-based content, such as descriptions, evaluations based on pre-defined cells, pictures, videos, and audio files [3]. In this context, the user fulfills two tasks: First of all, he creates (e.g. a text) or he selects (e.g. in case of an evaluation) [15]. It is possible to define UGC based on three characteristics [32]:

- Publication: The content has to be published online and made available for a broad public.
- Creative effort: The user has to have put in a certain degree of creativity to create content (dependent on the context).
- Creation outside of professional routines and practices: The user creates content outside of professional routines and practices.

2.2 Business Model

The term "business model" evolved around the year 2000 in connection with the new economy boom and came to be a much used expression. Astonishingly, a commonly accepted definition could neither in research nor in practice be established [21]. One definition in the broader sense interprets a business model as a description of how the business actually works [5]. Other definitions in literature differ according to the focus they address. With respect to *Timmers* [29] and *Heinrich* [8], we define the term "business model" as follows: A business model describes the architecture of a company at a specific point of time. It models all relevant dimensions of the company and explains how products and services are produced and profit is generated. Moreover, it describes the benefit, i.e. the value proposition, and service offering (products and services) for all internal and external actors, the actors themselves as well as the information flow and flow of goods and services.

Scientific literature deals with the subject of business models from two different perspectives. Some publications deal with the definition of the term or specific instances of business models [22]. Others focus on the classification of business models and the definition of typologies [e.g. 14, 29] or building blocks of business models to analyze existing models or propose new ones [e.g. 2]. *Wirtz* [30] and *Osterwalder* [18] suggested well-known and widely used approaches. *Wirtz* suggests the breakdown of the entire business model into partial models, such as the production or the profit model. *Osterwalder* devised a so called Business Model Ontology (BMO), which allows for a holistic modeling of the business model based on nine building blocks [18, 19]. The BMO has been derived from elements and models to be found in literature. The resulting building blocks are: value proposition, target customer, distribution channel, customer relationship, key resources, key activities, partnership, revenue model, and cost structure. These nine building blocks are well established for the description and analysis of business models [19]. Due to this and its holistic approach, the BMO is used for the following analysis of UGC business models.

2.3   Business Models for User-Generated Content

Business models can make use of UGC in many different ways (e.g. [9], [32]). It can, for example, be integrated into an existing business model and be used to support the original service offering. Amazon or ebay make use of this by letting the customers evaluate products and/or sellers to build a reputation and with that trust for other customers. Another possibility is to focus the business model on supporting the creation and publication of UGC for third-parties. This is, for example, done by Wordpress who offers users to run a blog and with that support their UGC process.

In this paper, we only refer to business models as UGC business models, if UGC is the primary service offering. Here, two types of so called intermediaries can be differentiated and with that two types of business models: On the one hand, we find intermediaries who offer market places [7]. They provide, according to the classical definition of a market place, a place where demand and supply meet. On the other hand, there are intermediaries, who serve as information suppliers by offering information and supporting their use [23].

3   Research Methodology

The research process in this paper is based on the multiple case study research as proposed by *Yin* [33] and *Eisenhardt* [6]. Multiple case study research belongs to the class of qualitative research methods and is used to analyze an object within its real context without being able to clearly distinguish between object and its environment. The cases studies analyzed in this paper aim at observing and explaining UGC business models. Multiple case study research represents, as the name suggests, a cross-section across multiple objects and compares as well as analyzes these objects at a specific point of time [26]. The selection of cases has to fulfill the requirement that they have to have a connection to the goal of research [6] [26]. We have analyzed six cases for our research: Ciao, HolidayCheck, iStockphoto, Qype, Tvype, and YouTube. The main selection criterion was that the business model has to correspond to the definition of a UGC business model as defined before, i.e. the use and support of UGC is the primary service offering. The selection has been done based on a thorough research of literature and the Internet. Scientific literature has been searched for relevant examples which show the implementation of UGC in practice. Moreover, non-scientific sources, such as online journals and weblogs, have been taken into consideration. Data collection was done based on document analysis and observations [33] and structured according to the BMO. Websites and the documents to be found there (company flyers, etc.) have been searched for relevant information. Online platforms as well as their functionality have been tested and evaluated by registration and participatory use. The information found for each case was assigned to the building blocks of the BMO, with the effect that for each building block a cross-section analysis could be made across all cases. The resulting building blocks have then been summed up within a morphological box [34]. Three steps lead to a morphological box [16]. First of all, the building block is further distinguished into different characteristics. Then, instances of these characteristics are assigned. The characteristics are listed in the first column and the respective instances are aligned to their characteristic in the rows. The third step refers to the use of the morphological box: If a business model is analyzed, the instances to be found for each characteristic are marked so that the overall instance of the business model is elicited.

4   Building Blocks of Business Models for UGC

4.1   Value Proposition

The first building block to be analyzed is the "value proposition". It takes into account the products and services as well as the benefits or value a company offers for the customers [18]. It has to be defined which needs are to be satisfied and with that, the focus of the service offering is set. Three different groups of stakeholders are addressed in the case of UGC business models: consumer (i.e. the consuming customer), author, and customers for adverts (i.e. advertisers). These groups of stakeholders are not strictly separated and a consumer can be an author at the same time (also refer to the definition of the building block "target customer"). We differentiate two different characteristics for the stakeholder group "consumer": "service offering (consumer)" and "type of content". For the stakeholder group "author", we also found two characteristics: "service offering (author)" and "benefit for the author". Finally, for the stakeholder group "advertisers" we found the characteristics "type of advert" and "benefit for the advertiser". Now, the different instances for the characteristics have to be deduced.

Both types of UGC business models base on an online platform. The access to the platform and its functionality is in both types the basic service offering. The consumer gains either access to information (e.g. Ciao for products, HolidayCheck for holiday destinations, or Qype for service providers) or can buy the right of use or property rights. In the case of iStockphoto, the consumer is interested in the photo as a resource, but not in the information the photo provides. Moreover, additional services are offered, such as the booking of hotel rooms when using HolidayCheck. As far as the type of content is concerned, the consumer has access to the typical types of UGC (e.g. text or video).

The basic service offering for the author is also the access to the platform or the market place, respectively, with the provided functionality. The author can either directly create content on the platform or upload externally created content, refine it, if need be, and can then publish it. YouTube, for example, allows for the upload of videos and provides a considerable amount of storage space for the author as well as a registration of the content in a directory. Moreover, the author can evaluate and comment on already existing videos. Qype, as another example, offers functionality to create content directly on the platform. Both the categorization of the content as well as the registration in a directory is also offered on this platform. Another service offering is the definition of a user profile which offers different possibilities of self-expression. In addition to the publication of personal data (e.g. e-mail address), the provided content is listed and communication as well as network opportunities with other users of this platform are offered. These platforms base on typical social networking applications with the focus on self-expression. Other platforms, such as HolidayCheck, aim at creating topic-driven communities with various alternatives to exchange opinions and experiences. The benefit for the author can basically be differentiated into the three instances "money", "glory", and "love" [15]. Intermediaries of market places often offer a financial share (money), information intermediaries can create a benefit by offering a certain status (e.g. expert traveler) (glory), and very often benefit is intrinsically created by the author himself, e.g. to be proud of the new design he proposed (love). A study conducted by *Stöckl et al.* [28], which analyzed the motivation of authors of UGC also proved this differentiation.

The stakeholder group "advertisers" can make use of two forms of advertising: typical online adverts with, e.g., banners, or affiliate marketing, where users are forwarded to the advertising company. Ciao, for example, offers both alternatives to its advertisers. Advertisers can "rent" space on the platform to promote products or the company. Promoted products can moreover be linked to the respective products in the online shop of the advertiser. Ciao integrates this form of advertising with a product comparison. YouTube offers its advertisers both, the possibility to rent advertising space on the platform as well as to integrate adverts into the contents provided by other users or allow for links to products. iStockphoto and Tvype do not especially cater for advertisers. The benefit for advertisers is the opportunity to directly address their target customers, which are a fairly large group and well segmented (e.g. according to the topic of the community or the self-expression of the users). Moreover, the advertisers might have a higher communication frequency with their customers based on affiliate marketing. Affiliate marketing can also provide a benefit for the consumer, as the case study Ciao shows. Many of Ciao's consumers using the product description and evaluations perceive the direct links to the products in connection with prices as an additional benefit, as they get more detailed information on the product. Table 1 presents the morphological box for the building block "value proposition" and the characteristics as well as instances derived from the case studies.

4.2 Target Customer

The building block "target customer" describes the group or rather groups of customers which gain the highest benefit and satisfaction of needs by the value proposition [18]. The stakeholder groups are the same as for the building block "value proposition": consumers, authors, and advertisers. As mentioned before, target customers can have more than one role within the different stakeholder groups and it might be necessary to adapt the form of communication. However, in the case of

Table 1 Value proposition "UGC Business Model"

Characteristic	Instances			
	Consumer			
Service offering (consumer)	Access to the platform	Information	Right of use or property rights	Intermediation
Type of content	Text	Evaluation	Picture	Video	Audio
	Author			
Service offering (author)	Access to the platform	Storage space; directory registration	Self-expression	Community
Benefits for the author	Money	Glory	Love	
	Advertiser			
Type of advert	Typical online adverts		Affiliate marketing	
Benefits for the advertiser	Opportunity to directly address target customers		Higher communication frequency	

the intermediaries of market places, such as iStockphoto or Tvype, the target customers do not take different roles. Consumers are only interested in the information provided and authors are normally only interested in the creation of content.

In contrast to that, Qype and YouTube show that target customers can be assigned to more than one stakeholder group. In the case of Qype, it is possible that authors provide evaluations and make an entry into a registry, if the object to be evaluated has not yet been registered. Moreover, advertisers can act as authors by making an entry for their company (i.e. advertiser = author). Users of YouTube can be both, consumer and author, because they do not only provide content, but also consume content of others. With this, the characteristic "role combination" can be jointly defined for all three stakeholder groups (cf. Table 2).

For the characteristic "customer relationship" the three stakeholders groups address three customer types: business (B), private consumers (C), and administration (A). The value proposition can be addressed to one or more of these customer types. iStockphoto and Tvype have mainly other companies (i.e. businesses) as target customers, for example media companies. Ciao or HolidayCheck do not restrict themselves to a specific customer type, but address mainly private consumers with the information offered. A specific focus on administration could not be found in any of the case studies. The definition of the target customer as well as a restriction to a specific group is based on the content that is to be provided, i.e. the topical focus of the information or the concentration on an industry. The target customers belonging to the group "author" are also diverse and there is no focus on one single customer type. The intermediaries for market places address private persons as well as freelancers and companies. The delimitation of the group of target customers is based on the competencies necessary for providing the content traded on the platform. iStockphoto, for example, requires professional media content in a certain quality. Other intermediaries, such as the information intermediary YouTube, do not have specific requirements with respect to quality and competencies; requirements with respect to content are defined by the file format and the respecting of basic rules, e.g. copyright. The target customers belonging to the stakeholder group "advertisers" are mainly companies, even though the service offering is open for any stakeholder group. The delimitation of the group of target customers is based on the content offered with respect to a specific topic, the community or the industry focus. It is possible to provide content with a broader focus and with that address advertisers from diverse industries (e.g. Ciao) or to focus on specific content and with that only address those advertisers having the fitting service offering. Qype offers a broad range of content for companies from different industries, but also allows for a focus on a certain sector: The advertizing space on the homepage addresses a broad range of customers. Moreover, there is also advertizing space offered for specific categories which only and specifically addresses customers fitting to the category.

Table 2 presents the morphological box for the building block "target customer" and the characteristics and instances derived from the case studies.

Table 2 Target customer "UGC Business Model"

Characteristic	Instances		
Role combination	None	Consumer = Author	Advertiser = Author
Consumer			
Customer relationship	B2C	B2B	B2A
Specific topic / industry focus	Yes	No	
Author			
Customer relationship	B2C	B2B	B2A
Competencies	Expert	Everybody	
Advertiser			
Customer relationship	B2C	B2B	B2A
Specific topic / industry focus	Yes	No	Partly

4.3 Distribution Channel

The building block "distribution channel" represents the connection between the value proposition and the target customer. It describes the different ways a company can establish contact with its customers to fulfill the service offering and communicate with them [18]. Both types of intermediaries, those offering market places as well as those offering information, only allow for an access via online platforms, i.e. through the medium Internet. The type of distribution channel is a direct sales or direct marketing channel. In addition, information intermediaries can also pursue indirect sales through other online media. YouTube, e.g., allows for the integration of content into other online media. It is, for example, possible that an online magazine integrates a YouTube video into a news article. The video is marked with the

YouTube logo which leads the reader to the YouTube platform. When watching the video, the reader gets the same ads as if watching it on YouTube. Even though the Internet is the uniform sales medium, it is possible to access the content through more ways than only the browser. Qype and Tvype, for example, offer their customers mobile applications for accessing their platforms. Table 3 shows the morphological box for the building block "distribution channel" and the characteristics as well as instances derived from the case studies.

Table 3 Distribution channel "UGC Business Model"

Characteristic	Instances	
Channel type	Direct	Indirect
Distribution medium	Internet	
Access	Browser	Mobile application

4.4 Customer Relationship

The building block "customer relationship" describes the type of relationship a company and a customer share. As a consequence, it serves the same purpose as the building block "distribution channel" and defines the connection between the value proposition and the target customer [18]. For the stakeholder groups "consumer" and "author", the relationship is defined based on "self services" or automated services provided by the online platform. The intermediaries provide the platform, but do not engage in customer care activities. Consumers and authors using the analyzed platforms in the case studies can access the service offering totally independently and can access, create and upload content. The cases of YouTube and Ciao show, how automated services are used to shape the customer relationship. Automated recommender systems suggest content which is supposed to be of additional interest for the customer, but has not directly been requested by him or her. Only general questions are supported by personal contact channels, e.g. email. Moreover, community tools, such as forums or accounts on social networks, are used for shaping the customer relationship. iStockphoto offers a forum for general discussions as well as for specialized discussions concerning questions and requests with respect to iStockphoto. Tvype uses the social network Facebook and the micro-blogging service Twitter for communicating with its customers, e.g. for receiving ideas for improvement and for answering questions. These forms of communication and building relationships form a customer relationship on a pure technical level.

For the stakeholder group of advertisers, the relationship can also be built on self service through the online platform. Qype offers its advertisers the possibility to calculate and book their preferred advertizing directly through the platform.

Advertisers often get a personalized customer care due to the different types and forms of advertizing. Ciao offers its advertisers detailed information on available forms of advertizing on the online platform. The consulting on the design of advertizing campaigns is done personally by Ciao employees. The customer relationships built here can either be technical or personal. Table 4 shows the characteristics and instances of the building block "customer relationship".

Table 4 Customer relationship "UGC Business Model"

Characteristic	Instances		
Consumer and Author			
Relationship type	Self service	Automated service	Community
Relationship level	Technical		
Advertiser			
Relationship type	Self service	Personal assistance	
Relationship level	Technical	Personal	

4.5   Key Resources

The building block "key resources" describes those resources which are of critical relevance for the production of the service offering [19]. Normally, key resources are differentiated into the categories "physical assets", "intangible items", "human resource", and "financial resources" [19]. The categories "physical assets" and "intangible items" have the highest relevance for UGC business models. The main key resource is the technical infrastructure, belonging to the physical assets. Almost all analyzed business models are dependent on the online platform and its availability as well as functionality. Intangible items also play an important role. They are the existing content as well as the base of authors and consumers. Ciao and Qype are dependent on an extensive content base to be attractive for their customers. This means at the same time to entertain a large base of authors for the creation of content. In contrast to that, Tvype needs a large network of consumers due to the strict delimitation of the addressed target customers. Moreover, the brand has a high relevance as an immaterial good (intangible item) for UGC business models. This can be explained using YouTube as an example: YouTube is a brand which has been influenced by Google during its development; this supports its differentiation from competitors like MyVideo or Clipfish. The category "human resource" does only play a subordinate role. The production of the service offering is only marginally dependent on highly specialized employees, because the content is provided by the crowd, i.e. the authors. Financial resources also are of lower importance since most of the business models normally do not need financial guarantees or high credit lines. Table 5 presents the two characteristics and the respective instances for the building block "key resources".

Table 5 Key resources "UGC Business Model"

Characteristic	Instances			
Resource category	Physical assets	Intangible items	Human resources	Financial resources
Important resource	Technical infrastructure; Online-platform	Content base	Consumer and author base	Brand
				Other

4.6 Key Activities

The building block "key activities" describes those activities which are of relevance for the implementation of the business model [19]. Basing on the explanation with respect to key resources, it can be said that a key activity of UGC business models is the operation of the online platform. The value proposition as well as the service offering can only and exclusively be provided by the platform. The operation of the platform in connection with its maintenance and development is one of the most important activities of this type of business model. In addition to this, the design of the platform with respect to the requirements of the stakeholder groups (e.g. functionality or usability) plays an important role. Moreover, marketing and sales are important activities. The attractiveness for the target customers for this type of online platforms is dependent on net effects [12]. This also has an impact on the key resources content base as well as base of consumers and authors. The online platform of HolidayCheck is only interesting for the consumer of content, if he actually gets the evaluation for the hotel he wants to book. The authors at Tvype are also only interested in publishing their content on the platform, if enough consumers visit the platform and with that demand exists for their work. And advertisers are interested in an as high as possible number of users which take interest in the adverts. In all UGC business models, the content is the main part of the service offering; that is why quality control is another important activity. Quality control at Ciao is, for example, done by the consumers and other authors. HolidayCheck or iStockphoto coordinate their quality control through employees. Table 6 shows the characteristics and instances of the building block "key activities".

Table 6 Key activities "UGC Business Model"

Characteristic	Instances			
Important activity	Maintenance and development	Marketing and sales	Quality control	Other
Target group for marketing and sales	Author	Consumer	Advertiser	
Quality control	Employee	Author	Consumer	

4.7 Partnership

The building block "partnership" describes the network activities within a business model. If key resources or key activities are sourced through external partners, this building block explains the existing partners, the relationships between the partners and their role in the value adding [18]. The design of the partner network considerably varies in the analyzed case studies. The following partnerships could be observed: partnerships with content providers, advertisers, marketers of advertizing space or other content providers. The type of partnership also varies: HolidayCheck established a classical demand-supplier-relationship with travel agencies and sells their products through the platform. iStockphoto implemented strategic partnerships with companies who have a connection to the content provided on the platform (e.g. printer companies). In general, partnerships are built for the sourcing and distribution of content. Sourcing of content is done through partners coming from the stakeholder groups "author" and "advertiser". The two cases Ciao and Qype show that content, i.e. information on products and companies, is offset with the return service "advertizing" (linking/presentation). With that, advertisers are systematically used as content providers. YouTube promotes a so called "content partnership" and praises the reach of the platform and the related advantages for advertizing purposes. YouTube establishes partnerships with private authors, who regularly provide content which is used by many consumers, in addition to partnerships with advertisers and companies. Authors accepting a partnership with YouTube participate in the advertizing revenues. Authors without a partnership status do not participate in the revenues. iStockphoto pursues a similar approach: Authors who regularly provide content are offered partnerships.

The distribution of content and the promotion of the platform can also be pursued through partnerships. HolidayCheck offers partner programs for other content providers. These providers can integrate content of HolidayCheck into their platforms and get a financial reward if the user is from there forwarded to Holiday-Check. Advertisers and authors normally also have the opportunity to use other content and support sales by that. Qype offers another form of partnership: Qype uses a partner for the marketing of advertizing space on their platform. Advertizing customers of Qype are thus completely attended to by the marketing partner and not by Qype itself. In addition to these already mentioned types of partnerships, there are financial partnerships with investors, who support young businesses in the early stages of formation. Table 7 shows the characteristics and instances for the building block "partnership" which could be derived from the case studies.

4.8 Revenue Model

The building block "revenue model" describes how the company generates revenues. There can be different sources of revenue [18]. Revenues can either be generated being dependent on transactions or being independent from transactions.

Table 7 Partnership "UGC Business Model"

Characteristic	Instances							
Type of partner-ship	Product partner	Content partner	Distribu-tion partner (Content)	Adverti-sing space marketers	Strategic partner	Financial partner	Other	None
Content partner	Author		Advertise customer		Other		None	
Distribution partner	Content provider	Advertise customer		Author		Other	None	

Moreover, they can be generated directly or indirectly through non-core business [31]. Intermediaries of market places normally choose a direct and transaction-dependent revenue generation by participation in the revenues coming from the UGC. Depending on the specific design of the revenue model, either the author receives a fee from the intermediary of the market place or the intermediary receives a fee from the author.

Information intermediaries generate revenues through advertising. Revenue is on the one hand generated indirectly and in a transaction-independent way by providing advertising space on the platform. It can also be generated directly and in a transaction-dependent way by forwarding to products and services of advertisers. If not only UGC is provided but products are placed and sold, revenues are generated directly and in a transaction-dependent way in the form of fees. Revenues are generated by intermediaries of market places and those of information through consumers and advertisers. Author can normally use the platforms free of cost. Table 8 presents the morphological box for the building block "revenue model" and the characteristics and instances derived from the case studies.

Table 8 Revenue model "UGC Business Model"

Characteristic	Instances			
Revenue source	Consumer		Advertiser	
Revenue type	Transaction-de-pendent: direct	Transaction-de-pendent: indirect	Transaction-inde-pendent: direct	Transaction-inde-pendent: indirect
Revenue stream	Sale (UGC)	Commission (Intermedia-tion)	Advert (typical online advert)	Advert (affiliate marketing) Other

4.9 Cost Structure

The building block "cost structure" describes which costs occur in connection with the production of the service offering [18]. We differentiate between cost-driven business models, which try to keep costs as low as possible and value-driven business models, which focus on the best possible fulfillment of the value proposition [19]. Moreover, we find business models with a high degree of fixed costs and those with a high degree of variable costs. The costs occurring in UGC business models primarily stem from the key activities and the key resources; these can either represent fixed or variable cost, depending on the implementation of the business model. The type of UGC also influences the cost structure. Requirements with respect to the technical infrastructure are much higher as regards video clips in comparison to mere text-based content. An important cost driver can be quality control. Costs are higher if it is done by internal staff compared to automated control by the author. Table 9 shows the characteristics and instances of the ninth and last building block "cost structure".

Table 9 Cost structure "UGC Business Model"

Characteristic	Instances						
Class of cost structure	Cost-driven			Value-driven			
Cost characteristic	Fixed costs			Variable costs			
Costs	Technical infras-tructure	Building / facilities	Workforce	Marketing and sales	Quality control	Commissions (authors etc.)	Other

5 An Example

The following example of YouTube shows how a business model for UGC can be analyzed based on the single morphological boxes. In combination the single tables result in a holistic reference framework for business models for UGC.

Table 10 Business model of YouTube

Characteristic	Instances			
Value Proposition				
Consumer				
Service offering (consumer)	Access to the plat-form	Information	Right of use or property rights	Intermediation
Type of content	Text	Evaluation	Picture	Video	Audio
Author				
Service offering (author)	Access to the plat-form	Storage space; di-rectory registration	Self-expression	Community
Benefits for the author	Money	Glory		Love
Advertiser				
Type of advert	Typical online adverts		Affiliate marketing	
Benefits for the advertiser	Opportunity to directly address target customers		Higher communication frequency	

Table 10 *(continued)*

Target Customer			
Role combination	None	Consumer = Author	Advertiser = Author
Consumer			
Customer relationship	B2C	B2B	B2A
Specific topic / industry focus	Yes		No
Author			
Customer relationship	B2C	B2B	B2A
Competencies	Expert		Everybody
Advertiser			
Customer relationship	B2C	B2B	B2A
Specific topic / industry focus	Yes	No	Partly
Distribution Channel			
Channel type	Direct		Indirect
Distribution medium	Internet		
Access	Browser		Mobile application
Customer Relationship			
Consumer and Author			
Relationship type	Self service	Automated service	Community
Relationship level	Technical		
Advertiser			
Relationship type	Self service		Personal assistance
Relationship level	Technical		Personal
Key Resources			
Resource category	Physical assets	Intangible items	Human resources	Financial resources
Important resource	Technical infrastructure; Online-platform	Content base	Consumer and author base	Brand	Other
Key Activities			
Important activity	Maintenance and development	Marketing and sales	Quality control	Other
Target group for marketing and sales	Author	Consumer	Advertiser
Quality control	Employee	Author	Consumer

Table 10 (*continued*)

Partnership							
Type of partnership	Product partner	Content partner	Distribution partner (Content)	Advertising space marketers	Strategic partner	Financial partner	Other / None
Content partner	Advertiser		Author		Other		None
Distribution partner	Content provider	Advertiser		Author		Other	None
Revenue Model							
Revenue source	Consumer				Advertiser		
Revenue type	Transaction-dependent: direct		Transaction-dependent: indirect		Transaction-independent: direct		Transaction-independent: indirect
Revenue stream	Sale (UGC)	Commission (Intermediation)		Advert (typical online advert)	Advert (affiliate marketing)		Other
Cost Structure							
Class of cost structure	Cost-driven				Value-driven		
Cost characteristic	Fixed costs				Variable costs		
Costs	Technical infrastructure	Building / facilities	Work-force	Quality control	Marketing and sales	Commissions (authors etc.)	Other

6 Conclusion

The sharing of knowledge and creativity as well as the participation of the many in the design and evaluation of products and services is a very important development regarding today's value creation. *Boyd* [4] stated in an interview: "Once companies realize the potential of social networking, we will see a revolution." As a consequence, it is time to analyze the mechanisms of creating value starting with researching the basis for that business models. This paper presents characteristics and instances of nine building blocks for UGC business models that finally result in a reference framework. Based on the BMO of *Osterwalder* [18] [19] these nine building blocks are value proposition, target customer, distribution channel, customer relationship, key activities, key resources, partnership, revenue model and cost structure, which allow for a holistic representation of business models. The presented building blocks and their characteristics and instances permit a classification of existing UGC business models and based on that a comparison of different UGC business models. Companies thinking about integrating this kind of business model can also develop, enhance or evaluate their already existing business model. Furthermore, entrepreneurs gain an overview and a design concept to implement such business models.

The presented building blocks are dependent on each other and as a consequence correlate. These correlations could only be addressed superficially. Further research has to be done to have a better understanding of these interdependencies and to make more extensive conclusions regarding this business model type. Case study research always results only in theoretical propositions and merely offers an analytical generalization [33]. Consequently, a further step could be an empirical validation of the presented characteristics and instances. Overall, the paper proposes a comprehensive reference framework for business models for UGC which can be used to study this business model type and to create new business models of this type for value creation.

References

1. Alby, T.: Web 2.0: Konzepte, Anwendungen, Technologien, 3rd edn. Hanser, München (2008)
2. Alt, R., Zimmermann, H.D.: Preface: Introduction to Special Section - Business Models. Electronic Markets 11(1), 3–9 (2001)
3. Bowman, S., Willis, C.: We Media - How audience are shaping the future of news and information. The Media Center at The American Press Institute, Reston (2003)
4. Boyd, S.: Interview: Facebook ist eine vorübergehende Phase. Frankfurter Allgemeine Zeitung (May 18, 2010)
5. Casadesus-Masanell, R., Ricart, J.E.: Competing Through Business Models. Working Paper Nr. 713. IESE Business School - University of Navarra, Barcelona (2007)
6. Eisenhardt, K.M.: Building Theory from Case Study Research. Academy of Management Review 14(4), 532–550 (1989)
7. Giaglis, G.M., Klein, S., O'Keefe, R.M.: The role of intermediaries in electronic marketplaces: developing a contingency model. Information Systems Journal 12(3), 231–246 (2002)
8. Heinrich, B.: Das Geschäftsmodell als Instrument zur Positionierung eines Unternehmens. In: Leist, S., Winter, R. (eds.) Retail Banking im Informationszeitalter - Integrierte Gestaltung der Geschäfts-, Prozess- und Applikationsebene. Springer, Berlin (2002)
9. Hoegg, R., Martignoni, R., Meckel, M., Stanoevska-Slabeva, K.: Overview of Business Models for Web 2.0 Communities. In: Proceedings of GeNeMe 2006, Dresden (2006)
10. Ickler, H., Schülke, S., Wilfling, S., Baumöl, U.: New Challenges in E-Commerce: How Social Commerce Influences the Customer Process. In: Ickler, H. (ed.) Proceedings of the 5th National Conference on Computing and Information Technology, NCCIT 2009, Bangkok, pp. 51–57 (2009)
11. Ickler, H.: An approach for the visual representation of business models that integrate web-based collective intelligence into value creation. In: Bastiaens, T., Baumöl, U., Krämer, B. (eds.) On Collective Intelligence. AISC, vol. 76, pp. 25–35. Springer, Heidelberg (2010)
12. Kollmann, T., Stöckmann, C.: Oszillationseffekte für Web 2.0-Plattformen - Kritische Masse Probleme im virtuellen Wettbewerb. In: Kollmann, T., Häsel, M. (eds.) Web 2.0 - Trends und Technologien im Kontext der Net Economy, Gabler, Wiesbaden, pp. 207–224 (2007)
13. Leimeister, J.M.: Collective Intelligence. Business & Information Systems Engineering 2(4), 245–248 (2010)
14. Linder, J., Cantrell, S.: Changing Business Models: Surveying the Landscape. Accenture Institute for Strategic Change, Cambridge (2000)
15. Malone, T.W., Laubacher, R., Dellarocas, C.: The Collective Intelligence Genome. MIT Sloan Management Review 51(3), 21–31 (2010)
16. Müller-Merbach, H.: The Use of Morphological Techniques for OR-Approaches to Problems. In: Operations Research 1975, pp. 127–139. North-Holland Publishing, Amsterdam (1976)
17. O'Reilly, T.: What is Web 2.0 (September 30, 2005)
   http://www.oreillynet.com/pub/a/oreilly/tim/news/2005/09/30/what-is-web-20.html (accessed September 15, 2010)
18. Osterwalder, A.: The Business Model Ontology - A Proposition in a Design Science Approach. Dissertation, University of Lausanne (2004)

19. Osterwalder, A., Pigneur, Y.: Business model generation: a handbook for visionaries, game changers, and challengers. John Wiley & Sons, New Jersey (2010)
20. Pousttchi, K., Schiessler, M., Wiedemann, D.G.: Analyzing the Elements of the Business Model for Mobile Payment Service Provision. In: Proceedings of the International Conference on the Management of Mobile Business (ICMB 2007). IEEE Computer Society, Washington, DC (2007)
21. Rentmeister, J., Klein, S.: Geschäftsmodelle in der New Economy. Das Wirtschaftsstudium (WISU) 30(3), 354–361 (2001)
22. Rentmeister, J., Klein, S.: Geschäftsmodell - Ein Modebegriff auf der Waagschale. ZfB-Ergänzungsheft 1, 17–30 (2003)
23. Rose, F.: The Economics. Concept, and Design of Information Intermediaries: A Theoretic Approach. Information Age Economy Series. Physica, Heidelberg (1999)
24. Sharma, S., Gutiérrez, J.A.: An evaluation framework for viable business models for m-commerce in the information technology sector. Electronic Markets 20(1), 33–52 (2010)
25. Skiera, B.: Preisdifferenzierung. In: Albers, S., Clement, M., Peters, K. (eds.) Marketing mit interaktiven Medien, Frankfurt, pp. 283–296 (1999)
26. Stake, R.E.: The Art of Case Study Research. Sage, Thousand Oaks (1995)
27. Stöckl, R., Grau, C., Hess, T.: User Generated Content. MedienWirtschaft - Zeitschrift für Medienmanagement und Kommunikationsönomie 3(4), 46–50 (2006)
28. Stöckl, P., Rohrmeier, T., Hess, T.: Why Customers Produce User Generated Content. In: Hass, B.H., Kilian, T., Walsh, G. (eds.) Web 2.0 - Neue Perspektiven für Marketing und Medien, Berlin (2008)
29. Timmers, P.: Business Models for Electronic Markets. Electronic Markets 8(2), 3–8 (1998)
30. Wirtz, B.W.: Business Model Management: Design - Instrumente - Erfolgsfaktoren von Geschäftsmodellen. Gabler, Wiesbaden (2010)
31. Wirtz, B.W.: Electronic Business, 3rd edn. Gabler, Wiesbaden (2010)
32. Wunsch-Vincent, S., Vickery, G.: Participative Web: User-created Content. Organisation for Economic Co-operation and Development (OECD), DSTI/ICCP/IE(2006)7/FINAL, JT03225396, Paris (2007)
33. Yin, R.K.: Case Study Research - Design and Methods, 4th edn. Sage, Thousand Oaks (2009)
34. Zwicky, F.: Entdecken, Erfinden, Forschen im Morphologischen Weltbild. Knaur-Droemer, München (1966)

Collective Intelligence Model: How to Describe Collective Intelligence

Sandro Georgi and Reinhard Jung

Abstract. A large number of scientific research exists, describing forms of collective intelligence (e.g. Wikipedia). But there are only few publications that describe how different forms of collective intelligence be described in general. In this paper, we therefore describe an approach how to characterise different forms of collective intelligence. We draw from existing research and build a comprehensive model and identify further characteristics to describe collective intelligence in a fine-grained manner. We propose a model with different characteristics, like form of cooperation, organisational pattern, and decision making process, which distinctively describe forms of collective intelligence and suggest possible attribute values.

Keywords: Collective Intelligence, Model

1 Introduction

Groups of individuals acting together to achieve better and superior results are not a new phenomenon. People have always teamed up since humans exist (e.g. for chasing deer, waging war, producing goods). What distinguishes collective intelligence from such common forms of cooperation is that information technology plays an important role in enabling the desired outcome. Without technology facilitating communication and coordination among participants, it would not be possible for large groups (such as for instance all the authors of Wikipedia) to coordinate their effort and disseminate information in such a fast way as it happens today. During the approximately last five years a significant shift in people's behavior of how they use the internet has taken place. This is due to new forms of communication and information exchange. People have created a taxonomy to tag (called folksonomy)

Sandro Georgi · Reinhard Jung
Institute of Information Management, University of St. Gallen, Mueller-Friedberg-Strasse 8, 9000 St. Gallen, Switzerland
e-mail: {Sandro.Georgi,Reinhard.Jung}@UNISG.CH

J. Altmann et al. (Eds.): Advances in Collective Intelligence 2011, AISC 113, pp. 53–64.
springerlink.com   © Springer-Verlag Berlin Heidelberg 2012

the content of pictures which they upload to Flickr [2] or similar platforms, built knowledge bases such as Wikipedia [28] or created t-shirt designs and share them on platforms like Threadless [26], where the best designs win to be printed and sold. By doing so, group participants act collectively and achieve results, which would not have been possible before. This is reflected in the definition of Malone et al. of the MIT Center for Collective Intelligence who define collective intelligence as "people and computers who are connected and act collectively more intelligently than individuals, groups, or computers have ever done before" [21].

A lot of companies apply collective intelligence in a more or less successful way to solve problems, but many are still not able to do so systematically. This makes the outcome/results of using collective intelligence up to a certain degree unpredictable. In order to successfully apply the principles of collective intelligence the way platforms such as InnoCentive [13] or GalaxyZoo [4] do, it is important to know the relevant characteristics which describe the various types of collective intelligence. A model which contains those characteristics will help invoking collective intelligence successfully in the future. Our long term research objective therefore is to build a situational method (the design artefact of the design science research approach as described by Hevner et al., [9]) which allows invoking collective intelligence based on specific requirements.

In the field of collective intelligence, research found is either about very specific examples of collective intelligence (such as Crowdsourcing or Wikinomics) or about collective intelligence in general. In what regards publications, the vast majority is about specific examples of collective intelligence while only little research is conducted about how to describe collective intelligence in general. The aim of this paper is to make a contribution to the latter. Drawing on existing knowledge, different research papers that propose models of how to describe collective intelligence are evaluated and then integrated into a more comprehensive model. Furthermore, additional characteristics, identified as relevant for the description of collective intelligence, are derived from system theory, process and task management.

The present paper is structured as follows: in the subsequent second part related research (i.e. existing approaches describing collective intelligence in general terms) is discussed and integrated into one large model. In section three further characteristics describing collective intelligence are identified and inserted/incorporated into the model derived in the second part. In chapter four a conclusion and outlook on further research is presented.

2 Related Work and a Synthesis of Existing Models

We identified only few contributions describing collective intelligence models. Malone et al. [19, 20] describe collective intelligence by means of the collective intelligence genome (see 2.2), Bonabeau [1] identifies collective intelligence approaches to mitigate biases in decision task (see 2.3), and Lykourentzou et al. [18] describe collective intelligence from system theory point of view (see 2.4).

In paragraph 2.1 of this chapter the existing definitions are discussed and we present the definition of collective intelligence which will be used in this paper. In paragraphs 2.2 to 2.4 the above mentioned contributions are discussed while in paragraph 2.5 further contributions are summarized. In paragraph 2.6 we present the synthesis of the existing models.

2.1 Defining Collective Intelligence

The research question of Malone et al. is "How can people and computers be connected so that – collectively – they act more intelligently than any individuals, groups, or computers have ever done before?" [21]. Lykourentzou et al. define a collective intelligence system as a "system which hosts an adequately large group of people, who act for their individual benefit, but whose group actions aim at and may result – through technology facilitation – in a higher-level intelligence and benefit of the community" [18]. Both research teams stress the fact that computers and technology respectively are part of collective intelligence. In addition to Malone et al., Lykourentzou et al. include the existence of individuals' and groups' objectives into their definition.

Bonabeau, surprisingly, forgoes the opportunity of stating his own definition of collective intelligence. He looks at collective intelligence with regard to making better decisions and, comparable to the other two research groups, he stresses the fact that recent technologies such as Web 2.0 applications enable companies to "tap into 'the collective' on a greater scale than ever before" [1].

We define collective intelligence similarly to Malone et al. but expand the definition by people's goal to gain a benefit from the participation: *People interact more intelligently (e.g. supported by information technology such as computers) than this has ever been done before. Thus, they achieve superior results than would have been possible by individuals alone. While contributing to the main goal by participating people pursue their own goals.*

2.2 The Collective Intelligence Genome

Malone et al. speak of "CI building blocks or 'genes' [that] can be recombined to create the right kind of system" [19, 20]. They have identified four dimensions which need to be described in order to describe collective intelligence: *what* is *how* done by *whom* and *why* people are doing it. The description of each of these dimensions is called a gene, which is part of the genome representing a specific form of collective intelligence [19, 20]. The different genes have various values, which we briefly summarise here:

- What-Gene: Malone et al. identified two different main tasks. People either *create* something (like t-shirt designs at threadless.com) or people *decide* about something (which t-shirt designs should be printed).

- Who-Gene: It is either a *crowd* who fulfils the task, or if the criteria for the crow are not met, it is a *hierarchy (management)* who is responsible.
- Why-Gene: People contribute or participate in collective intelligence to gain either money (financial reward), love (intrinsic motivation) or glory (recognition of one's achievements; often found in open source programming communities).
- How-Gene: this gene is dependent on *what* is being done:

  - *Creation* can be carried out in the form of a collection (the task can be split into items which can be solved independently of each other), a contest (subtype of collection with competing people), or a collaboration (there exist dependences between the various subtasks of the main task).
  - *Decisions* can be made either as a *group (voting, consensus, averaging* or *prediction market)* or by *individuals*.

2.3 Mitigating Biases in Decision Tasks

Bonabeau identifies two distinct basic tasks for which collective intelligence can be used: *generating solutions* and *evaluating different alternatives* [1] which are similar to the tasks identified by Malone et al.: *create* and *decide*. According to Bonabeau, each of the above mentioned tasks is subject to various biases (e.g. self-serving bias, pattern obsession etc.) which can be overcome by tapping into the collective intelligence of people. In order to achieve this, he proposes three different approaches [1] which he calls 'outreach' ("tap into people or groups that have not traditionally been included"), 'additive aggregation' ("collect information from myriad sources and then perform some kind of averaging"), and 'self-organisation' ("mechanisms that enable interactions among group members can result in the whole being more than the sum of its parts"). He furthermore identified various *key issues* which need to be thought of, such as the decision process (*decentralized decisions* vs. *distributed decisions*), *diversity* vs. *expertise* of individuals of the crowd, and *engagement* which refers to people's motivation to participate: *cash rewards, recognition,* and *the desire to transfer knowledge or share experiences*.

2.4 The Collective Intelligence System

Lykourentzou et al. provide a model for the description of collective intelligence systems [18] which consists of a categorization of active (crowd behavior does not pre-exist but is created and coordinated through specific system requests) and passive systems (individuals act as they would normally do without the system's presence). The latter systems can be subdivided into one of following three kinds: collaboration, competition, or a hybrid type. Lykourentzou et al.'s model further contains a collective intelligence model which includes certain attributes such as sets of actions, system state and the objectives of the participants as well as three important functions (user action function, future system state function and objective function). They additionally identify important factors of influence which need to be considered such as the critical mass of the system and the participants' motivation (*monetary compensation, intrinsic* motivation like *self-fulfilment motivator*, and *social recognition*).

2.5   Further Contributions

Each of the following three sources describes a specific approach about how collective intelligence can be utilized in a very specific form. Despite the fact that each approach in its original publication is described in rather general terms, the descriptions are much more specific (even too specific for our purposes) than the above mentioned three models and are therefore also of some importance for the field and research of collective intelligence. Thus, the respective approach is mentioned and briefly described hereafter but not incorporated into our collective intelligence model.

In their 2006 publication Tapscott and Williams describe what they call Wikinomics and how mass collaboration can change everything [25]. They identify four main aspects which are important for Wikinomics: being open, sharing, peering, and acting globally. In his famous wired.com article Howe proposes a form of mass collaboration which he calls Crowdsourcing [11, 12]. Tasks which were formerly executed by the company itself shall be published to a large audience which then can turn in possible solutions. Crowdsourcing is related to outsourcing [22] and is partially based on its values [23] but differs insofar as tasks are not outsourced to known companies but to a group of unknown people called the crowd. In his work on the wisdom of crowds Surowiecki identifies [24] four main aspects which need to be met in order to form a wise crowd: diversity of people, independence of people's opinions, decentralisation, and aggregation of the individual's judgments into a collective decision.

2.6   Synthesis of the Three Models

The three models presented in paragraphs 2.2 to 2.4 share some similarities but each model in its turn also contributes new, unrelated characteristics to describe collective intelligence. We deem the existing models to be of very good quality already, but think that they need to be expanded in order to describe collective intelligence in a more accurate way. Our model and its characteristics with corresponding values derived there from is summarized in table 1.

Our model differentiates between individuals' and goals of the community while the basic collective intelligence form can either be passive or active. Crowds or hierarchies (which are formed of professionals and/or amateurs) characterise the organisational pattern and cooperation can take place as collaboration, competition, collection or hybrid types. To use collective intelligence outreach, aggregation or self-organisation can be used as approaches. Groups know other decision making

Table 1 Collective intelligence model with characteristics and values

Goal	
Individuals	Depends on each individual
Community	Create, Decide
Basic Collective Intelligence Form	
Passive System, Active System	
Organisational Pattern	
Crowd, Hierarchy	
Form of Cooperation	
Collaboration, Competition, Collection (as a special case of the competition), Hybrid Type of competing groups with collaborating individuals	
Approach to utilise collective intelligence	
Outreach, Aggregation, Self-Organisation	
Professional Background of Individuals	
Experts, Amateurs	
Decision Making Process	
Groups	Voting, Consensus, Averaging, Prediction Markets
Individuals	Markets, Social Networks, Final Ballot
Form of the Decision-Making Process	
Distributed, Decentralised	

forms than individuals do while the decision-making process takes part either distributed or decentralised. Various reasons motivate people to participate: money, fame/glory, recognition, love and/or the desire to share knowledge.

The new and extensive model we derived from the existing models needs to be enhanced further in order to unambiguously describe collective intelligence. In the following chapter we identify additional characteristics which need to be taken into account when describing collective intelligence.

3   Characteristics Describing Collective Intelligence

As mentioned above, the existing models used to describe collective intelligence are already of a very good quality but describe the phenomenon still in very generic terms. In this chapter we will present further characteristics which lead to a better understanding of collective intelligence.

In paragraph 3.1 of this chapter, we discuss the objective of a task, look at the size of a contribution needed for collective intelligence in paragraph 3.2, the input and output of the process are discussed in paragraphs 3.3. and 3.4 and identify the various stakeholders involved in using collective intelligence in paragraph 3.5. The chapter closes with a synopsis of the newly identified characteristics describing collective intelligence.

3.1 Objective of a Task

According to Hoffmann [10] in organization theory a task can be characterised by an objective which needs to be achieved through a physical or mental target performance. Each task can, therefore, be paraphrased as "the objective to be achieved", "what needs to be done to achieve the objective", "how is it achieved", and "who is doing it". The just mentioned constituents "*what* needs to be done", "*how* it is achieved" and "*who* is doing it" are already sufficiently described by the model presented in chapter 2. The objective of the task, i.e. "what is to be done", is also described but only in a very broad way: *create* or *decide*. We argue that there is a need for a more fine-grained description of the objective of a task or rather the objective which shall be achieved through the use of collective intelligence. *Creating* and *deciding* may serve as main categories comprising several sub-categories like creation of *knowledge*, (product) *design/descriptions*, or *physical objects* for creation (see also paragraph 3.4). , Decisions, on the other hand, can be made about the *correctness* of knowledge (Wikipedia), the *best* design/description (Threadless, InnoCentive) or the *most suitable* physical product (InnoCentive).

3.2 Size of Contribution

When looking at collective intelligence it becomes clear that in most cases people contribute (very) small parts to a greater goal. At Galaxy Zoo for instance, galaxies are identified among other characteristics by their form (completely round, in between, or cigar shaped) or by the number of their spiral arms. The final classification of a galaxy is accomplished through a number of votes expressed by individuals used to create the average of 39 distinct votes per galaxy [17] for the same galaxy), a behaviour we call averaging in the decision process. People contributing such small parts to a greater goal are named *clickworkers* after the homonymic NASA project in which craters on planets and asteroids were identified.

Wikipedia, as another example of collective intelligence, contains millions of articles while the average author contributes only to a small number of articles to this encyclopaedia. InnoCentive also makes use of a large mass of so-called solvers [14] who answer questions asked by companies. In contrast to the just mentioned examples, however, the problems to be solved range from rather small to large amounts of effort. This leads to the conclusion that the *size of the contribution* of each person varies according to the specific problem and the form of collective intelligence used to resolve it.

3.3 Form of Input

According to Johansson et al. [16] a process is "a set of linked activities that take an input and transform it to create an output. Ideally, the transformation that occurs in the process should add value to the input and create an output that is more useful and effective to the recipient either upstream or downstream." Hammer & Champy [8] define a process as "a collection of activities that takes one or more kinds of input and creates an output that is of value to the customer." When looking at the various platforms that use collective intelligence it can be noticed that different forms of input and output can be identified.

Threadless offers on its submit page a "TeeSubmissionKit" [27] with a template, design rules and example pictures, which can be summarized as some sort of "instruction". A similar instruction is offered at Wikipedia where a tutorial about how to become a Wikipedia contributor can be found [29]. In both cases no actual input is provided as only instructions are offered about how to achieve the desired output. Rather, the input is the imaginativeness for shirt designs (Threadless) resp. people's knowledge (Wikipedia). Admittedly, however, in the case of Wikipedia there can be some sort of input if the contributors do not initiate a new article but only edit an existing one. In the latter situation, the available knowledge serves as an input that needs to be ameliorated.

At InnoCentive the solvers get a challenge description as an input and are then asked to use their intellect to find a solution. At Galaxy Zoo clickworkers first receive some instructions about the nomenclature to be used in the form of a short tutorial [5] and then get displayed randomly chosen pictures of galaxies as an input. These pictures, in a similar way as the challenge descriptions of InnoCentive, can be considered as raw material which is then transformed into the desired output, i.e. denomination or the solution to the problem.

Summing up, it can be noted that in most cases some sort of instruction is provided which sometimes serves as an input (e.g. at Threadless) while in some cases other input is necessary like pictures etc.

3.4 Form of Output

Both Johansson et al. [16] as well as Hammer & Champy [8] include in their definitions that the result of a process is, through transformation/creation, an output. Collective intelligence can be used to generate various forms of output which need to be looked at more closely. In the case of Wikipedia the output produced by the editors is knowledge while at Threadless the result is the print designs of shirts (or pullovers, hoodies etc.), and at InnoCentive it may be both knowledge and product designs in the form of either descriptions or physical evidence [15]. While knowledge is a form of output which is intangible, a shirt design or a design for a product in general is (at least when it is printed/produced) a tangible form of output.

We argue that the type of the output of a task solved with collective intelligence varies according to the specific form of collective intelligence used and can either be tangible or intangible.

3.5 Stakeholder

According to ISO 10006 stakeholders of a project are all those people who are interested in or in some way affected by it. In Freeman's book on *Strategic Management. A Stakeholder Approach*, one of the most renowned contributions to the identification of stakeholders of a company, the term stakeholder is defined as "those groups without whose support the organization would cease to exist" [3] and it originally included shareowners, employees, customers, lenders and society.

When looking at the use of collective intelligence, various stakeholders can be identified. First of all, there is the group of people who wants to use collective intelligence to achieve a desired objective. In case of Threadless a stakeholder is the organization behind Threadless who above all wishes to earn money while at Galaxy Zoo the scientists, who need their data examined resp. galaxies classified, represent a stakeholder. At InnoCentive, comparably to Threadless, CEO Dwaine Spradlin and his company want to make money. With InnoCentive, however, the company is not just the "owner" or initiator of collective intelligence like in the other two cases but it also serves as a mediator for others, i.e. the seekers, to utilize collective intelligence for the problem solving. The seekers ultimately also wish to earn money but are not the initiators of the collective intelligence.

Another group of stakeholders which can be identified is the contributors. These contributors do the actual work resp. are actually the intelligence needed. They provide knowledge (Wikipedia), imagine designs (Threadless), come up with innovative solutions for challenges (InnoCentive), or do the hard work of classifying galaxies (although most people claim that it is pure fun watching the pictures and classifying galaxies that no one else might have seen before and that "couch potato universe surfing is just plain cool" [6]).

A third group of stakeholders are the actual *beneficiaries* of the use of collective intelligence. This group overlaps with the initiators, who benefit by achieving their objective of earning money. The beneficiaries, however, also consists of people who profit from the output of employing collective intelligence. They for instance have access to a free online encyclopaedia (whose quality is close to the encyclopaedia Britannica [7]), and they get stylish t-shirts (Threadless).

3.6 Synopsis of the Identified Characteristics

So far we have identified five new characteristics in chapter 3 which need to be considered when describing collective intelligence (see table 2): objectives of the task are creation of knowledge, design/description of products or building physical products or decisions about the correctness of knowledge, the best design/description or the most suitable physical product. The size of people's contributions ranges from small to large while input is instruction (manuals, tutorials etc.), intellect or raw material which results in tangible or intangible output. Stakeholders of collective intelligence are initiators (of the collective intelligence), contributors and beneficiaries.

Table 2 Characteristics describing collective intelligence

Characteristic	Possible Values
Objective of task	Creation of knowledge, designs / descriptions of products / services, physical products.
	Decision about the correctness of knowledge, the best design/description, the most suitable physical product.
Size of contribution	Ranges from small to large
Form of input	Instruction, Intellect, Raw Material
Form of output	Tangible Output, Intangible Output
Stakeholders	Initiators, Contributors, Beneficiaries

4 Conclusion and Outlook

It is not a recent invention that people collaborate and join their knowledge and workforce. For as long as humans have existed, they have collaborated and thus achieved more as a group than they would have been able just on their own. The boost which collective intelligence has witnessed in the last two decades mainly comes from newly available information technologies. They enable people to connect with each other very easily and to exchange information in very short periods of time.

Many contributions can be found that present fascinating ideas about how collective intelligence can be used and how specific systems look like. However, only few research papers are available which offer models for the description of collective intelligence in general terms.

In this paper we synthesized existing models based on the work of other research groups into a new and more comprehensive framework. This synthesized model already helps to describe collective intelligence with more accuracy but still needs to be expanded. We do so by identifying the additional characteristics of the objective of a task, the form of the input and output of the process, the size of the contribution, and the various stakeholders involved.

A category system needs as many characteristics as are necessary to unequivocally describe the categorized objects and the categories must be both collectively exhaustive and mutually exclusive. For the time being we accept the characteristics and their possible values identified by Malone et al., Bonabeau and Lykourentzou et al. as being the proper characteristics to describe collective intelligence. Whether this holds true or not for the future and whether more and/or other characteristics in addition to those presented in chapter 3 are needed to describe collective intelligence must be examined further. Furthermore we do not so far consider dependencies among the characteristics in our model. Further research must thus be conducted to establish more and above all the appropriate dependencies among the identified characteristics. In order to do so, we plan to identify and analyse a number of forms of collective intelligence and derive their unique characteristics. With enough forms described it should be possible to identify types of collective intelligence, i.e. clusters of forms with identical attribute values for the various characteristics, and therefore build categories. Based on this, we plan to build a situational method to systematically invoke collective intelligence to solve specific problems.

References

1. Bonabeau, E.: Decisions 2.0: The Power of Collective Intelligence. MIT Sloan Management Review 2, 45–52 (2009)
2. Flickr Foto Sharing, http://www.flickr.com
3. Freeman, R.E.: Strategic Management. A Stakeholder Approach. Pitman, Boston (1984)
4. Galaxy Zoo Hubble, http://www.galaxyzoo.org
5. Galaxy Zoo Hubble: How to take part,
   http://www.galaxyzoo.org/how_to_take_part
6. Galaxy Zoo Hubble Forum, What makes Galaxy Zoo interesting, About this board,
   http://www.galaxyzooforum.org/index.php?topic=68.0
7. Giles, J.: Internet encyclopaedias go head to head. Nature 438, 900–901 (2005)
8. Hammer, M., Champy, J.: Reengineering the Corporation: A Manifesto for Business Revolution. Harper Business, New York (1993)
9. Hevner, A., March, S., Park, J., Ram, S.: Design Science in Information Systems Research. MIS Quarterly 28, 75–105 (2004)
10. Hoffmann, F.: Aufgabe. In: Grochla, Erwin (eds.) Handwörterbuch der Organisation, Poeschl, Stuttgart, pp. 200–207 (1980)
11. Howe, J.: The Rise of Crowdsourcing. Wired 14.06 (2006),
    http://www.wired.com/wired/archive/14.06/crowds.html
12. Howe, J.: Crowdsourcing: How the power of the crowd is driving the future of business. Crown Business, New York (2008)
13. InnoCentive, where the world innovates,
    http://www.innocentive.com
14. InnoCentive Problem Solvers,
    http://www2.innocentive.com/problem-solvers (retrieved)
15. InnoCentive Innovation Challenge Types,
    http://www2.InnoCentive.com/problemsolvers/
    innovation-challenge-types (retrieved)
16. Johansson, H.J., McHugh, P., Pendlebury, A.J., Wheeler, W.A.: Business Process Reengineering: BreakPoint Strategies for Market Dominance. Wiley, New York (1993)
17. Lintott, C., Land, K., Slosar, A., Andreescu, D., Bamford, S., Murray, P., Nichol, R., Raddick, M.J., Schawinski, K., Szalay, A., Thomas, D., Vandenberg, J.: Galaxy Zoo: the large-scale spin statistics of spiral galaxies in the Sloan Digital Sky Survey. Monthly Notices of the Royal Astronomical Society 4, 1686–1692 (2008)
18. Lykourentzou, I., Vergados, D.J., Loumos, V.: Collective Intelligence System Engineering. In: Proceedings of the International Conference on Management of Emergent Digital EcoSystems, pp. 134–140 (2009)
19. Malone, T.W., Laubacher, R., Dellarocas, C.: Harnessing crowds: mapping the genome of collective intelligence. Working Paper, MIT Center for Collective Intelligence (2009),
    http://cci.mit.edu/publications/CCIwp2009-01.pdf
20. Malone, T.W., Laubacher, R., Dellarocas, C.: The Collective Intelligence Genome. MIT Sloan Management Review 2, 20–31 (2010)
21. MIT Center for Collective Intelligence: Research Overview,
    http://cci.mit.edu/research/index

22. Schenk, E., Guittard, C.: Crowdsourcing: What can be Outsourced to the Crowd, and Why? Working Paper, aboratoire de énie de la Conception - Institut National des Sciences Appliquées de Strasbourg, BETA - Bureau d'économie théorique et appliqué, Université de Strasbourg (2009),
    http://ideas.repec.org/p/hal/wpaper/halshs-00439256_v1.html#download
23. Storey, M.D.: Beyond the Lone Reverse Engineer: Insourcing, Outsourcing and Crowdsourcing. In: IEEE 16th Working Conference on Reverse Engineering, Lille, France (2009)
24. Surowiecki, J.: The wisdom of crowds. Anchor Books, New York (2005)
25. Tapscott, D., Williams, A.D.: Wikinomics: How mass collaboration changes everything, Atlantic, London (2008)
26. Threadless graphic t-shirt designs, http://www.threadless.com
27. Threadless Submit, http://www.threadless.com/submit
28. Wikipedia, http://www.wikipedia.com
29. Wikipedia editing tutorial – introduction,
    http://en.wikipedia.org/wiki/Wikipedia:Tutorial

The Participatory Roles Play Simulation in a Social and Collective Learning Context

Aurelie Aurilla Bechina and Tone Vold

Abstract. Knowledge building and behavioral change takes place by participation in collective learning process. Participatory learning methods and intellectual interaction between people are an effective part of professional development. This paper discusses the role of game play in engaging actively stakeholder in a collective learning. This paper intends to investigate the factors facilitating the learning capability while designing a participatory role play model process. In order to delineate a general model, we have conducted two experiments in different Norwegian organizations. *The goal was to investigate the evidences of using role play game in a participatory as a suitable tool fostering collective learning.*

Keywords: Social Game, Knowledge Acquisition, Collective Learning.

1 Introduction

The last decades, the use of personal computers, smart mobile device and the internet has changed the way people are learning and acquiring new Knowledge. The emergence of games and the growing number of players have pushed the entertainment industry to develop more and more educational games. Only recently, the academic world started to investigate how serious games and computer simulation games could enhance the learning processes [2, 19, 24]. Latest research works have demonstrated that people tend to learn more easily when they are entertained and are engaged in the creativity process involving thinking and emotion [5, 11]. For

Aurelie Aurilla Bechina
University College of Buskerud, HIBU, Kongsberg, Norway
e-mail: aurillaa@hihm.no

Tone Vold
University College of Hedmark, ØSIR, Rena , Norway
Norwegian University of Science and Technology, Trondheim, Norway
e-mail: tone.vold@hihm.no

J. Altmann et al. (Eds.): Advances in Collective Intelligence 2011, AISC 113, pp. 65–77.
springerlink.com   © Springer-Verlag Berlin Heidelberg 2012

example, the work done by Blunt concludes, the application of serious games significantly increases learning capability [3]. One reason that is often invoked is closely related to the concept of constructivist learning. In constructivism theory, the learner actively constructs knowledge and learning by integrating new information and experiences into what they have previously understood. The learner by building cognitive structure combines propositional knowledge often referred as fact, concepts and procedural knowledge. This type of knowledge is usually related to techniques, skills and abilities [16].

In addition, research studies have demonstrated that knowledge construction is best effective when taken place in a social context, and in a setting in which new knowledge and skills are being used. Furthermore, there is a general acknowledgement that people learn best through interaction with others [29].

Compared to the traditional learning environment, the game based learning involves the learner actively in the learning environment. They are several factors in favors of using games to educational practices or to foster knowledge acquisition [24]. One of them is the recognition that the play constitutes an important part of human cognitive and social development. Gee reported in his study of video games as a learning tool, that students are more likely to assimilate content by engaging themselves in a process of discovery [11].

Moreover, people tend to construct knowledge in a best way when they are engaged personally; therefore game for learning purpose should involve content-based scenario or purpose. Active participation in tasks is seen also as a requirement for an active learning process; in fact, it is crucial to not separate the acquired knowledge from its everyday usage.

It is important to be aware as stated by Schnotz [23], that cognitive processing is not the only factor contributing to effective learning, but affective impacts and motivational goals should be taken into account as well. Therefore, if educational games are more engaging and appealing to learners interacting actively with these learning environment, it is worth to investigate further use of these new game based learning approaches [4].

In the context of our research study, we focus mainly on the role play game simulation. There are plethora of studies highlighting the benefits of using Role Play Game (RPG) [17, 18] as an innovative way to teach and learn new skills or acquire knowledge.

RPG simulation is considered to be a good teaching approach by fostering Collective learning in functional groups. Collective learning contributes for groups or teams to acquire collectively new competences or knowledge that will help them to deal or to adapt to new business or social settings [1].

Playing role game simulation is suitable to create group dynamic with evolving relationship ties among the participants. It is recognized that playing RPG simulation in a business context can contribute to facilitate team development, improve communications effectiveness, disclose some behavioral attitude towards a specific issue and so forth [6].

Literature investigation indicates that using role play simulation for fostering learning capability is not new and that most of the time the game has a predefined learning objective described in the scenario. This experiential learning technique is generally used with predefined roles and role play scripts [20, 25]. However, in our research study, we intend to use an overall new approach by rather involving the learners in the development process of their own learning objectives. Participants are encouraged to define the roles and scripts/scenario based on their own learning objectives. This participatory view is based on several learning theorists that claim the importance of the personal involvement by the learner [7, 8, 13, 22]. We believe that the implication of the learners in defining the learning objectives should enhance their own experiential learning.

The paper discusses the concepts and challenges in using role play game simulation in order to foster user's participation in a collective learning effort. In order to validate our assumptions or hypothesis, we have conducted two experiments involving two Norwegian organizations. The goal was to investigate the evidences of using role play game in a participatory as a suitable tool fostering collective learning in a group of people. To this end, we have conducted an interactive workshop encompassing two sessions. During the first session, the participants were requested to define collectively some learning objectives, to design a script/scenario for the role play. The second session gave the opportunity to evaluate the process and the outcome of the first part of the session by using an action after review approach.

The next section discusses how the learning processes can take place both at individual and organizational level. A literature review based on collective learning, user participation and role play is outlined. The third part presents the context of study based on the experiences gathered from the involved organizations. This section describes the adopted methodology to conduct a role play session using a participatory approach. Based on literature reviews, data collected during the workshop a model encompassing factors facilitating the learning capability while using a participatory approach is presented in the section four.

2 Social and Collective Learning

The concept of collective social learning is seen as a process that occurs among a group of people in which shared meanings are constructed and who seeks to improve a common situation and take action collectively [12]. This concept of collective learning reflects the growing interest among adult learning theorists in developing alternatives learning theories to explain how and why learning occurs in groups [12]. Although several research studies highlight that cognitive dimension is an integral part of the rule and routines in a collective learning process, there are still others factors that need to be further investigated [15]. For example, active user participation in a learning process is considered as a positive approach. Eikeland and Berg claim that learning cannot be based on theoretical input only, but that there is a need for experiencing [10]. In the same line, the work of Dewey [7] demonstrates that learning by experiencing is essential; however, linking action and thought to the experience is even more fundamental in order to enhance the learning capability [7]. Silberman defines this process as an experiential learning reflecting the learner's involvement in some specific activities that enable them to "experience" what they are learning. In addition, experiential learning provides the opportunity to reflect on these activities [25]. The experiencing can range from simulating real life to working situations.

Another theorist that has influenced experiential learning is David A. Kolb. He states: *"Learning is the process whereby knowledge is created through the transformation of experience"* [14]. Knowledge about what they have learned is to a large extent what we want the participants of the experiments to display during the phase of idea and scenario building. This is reflected in the Kolb's model of experiential learning cycle that is depicted in figure 1.

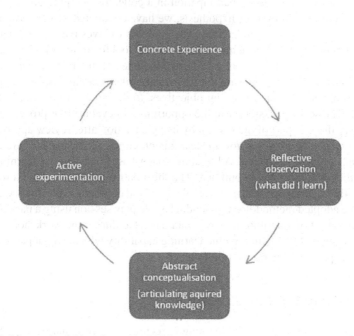

Fig. 1 Kolb's experiential learning cycle (1984)

To achieve learning from an experience, the knowledge obtained from the concrete experience must be contemplated on through reflective observations. An articulation can provide an abstract conceptualization that can be tested through active experimentations and gives new experiences to reflect upon. However, one major drawback of experiential learning is related to the lack of clarity in the learning outcomes. Furthermore, Academic institutions believe that this learning approach is quite restrictive since experiential learning could legitimize individual experience as the main source of knowledge and understanding [21]. Therefore, it is recognized that Social interaction and co-operation are important dimensions in the social learning. As stated by Wenger learners are social beings that are learning through developing competencies by taking part in activities where it is possible to derive meanings from [28]. Hence, taking into account the social dimension in experiential learning gives the participants a collective opportunity of learning from each others.

Amongst innovative learning approaches, Role play game simulation is considered as a good way to enhance experiential learning combined with social interaction between participants within a specific context that will simulate for example a business case situation. Participants can simulate a situation collectively and thus increase the collective intelligence that contributes to suggest new problems solving approaches.

Role play simulation is a game where social conflicts and group decision making are simulated. Traditionally, the topic and the roles are already predefined while the outcomes of the game are non-determinist. During the role play simulation games, learners are in a situation where decisions making is based on real or hypothetical model situations defined by a set of rules that govern their fictitious reality [26]. This approach is particularly suitable for social learning emphasizing not only knowledge acquisition but as well competences and communication skill building, behavioral attitude improvement and so forth. The learners through a simulated situation will learn in a safe, friendly and motivating environment.

Several research studies have demonstrated the benefit of using RPG simulation in various application domains. For example, Moroccan smallholder farmers have used a role play game together with a simulation tool in order to learn how to design drip irrigation systems. The researchers have designed and developed a role-playing game that was in first instance used to raise awareness among farmers about the scope and contents of a joint irrigation project and to know the knowledge level of the farmers. A policy simulation exercise based on the actual field situation enabled farmer groups to design their own joint drip irrigation project. The evaluation of the experiment demonstrated that a role-playing game provided valid learning experience while the realistic simulation supported concrete decision making [9].

Role play simulations have been as well used to help new custody officers to learn how to deal with different type of prisoners in custody ward [21]. The game have been developed using the knowledge of experienced custody officers, the scenario/script has been written by a professional writer. In order to conduct an assessment of the knowledge acquired by the participants, the role play session have been video recorded. It has been used during the debriefing process in order to discuss various decision taking, or attitude while playing the role play game. The learning outcomes have been defined as a preamble of the role play game. This is quite often the case in most of the research studies that we investigated.

In our approach, we took a step further by including the participants themselves in the design of the scenario or the script and in the role specifications. We assumed that the participatory process should contribute to secure a collective understanding of the learning outcomes that will be defined during the session. The participatory learning has been widely discussed in the literature as an interesting approach to professional skill development in order to enhance learning to improve one's practice. The argument for this approach is related to the belief that continuous professional development is not only engaging the learner in the process; but it also includes learning from others.

Although few research studies have been conducted in order to evaluate the impact of active participation in a learning process, there is still too little research done on the participatory process implication in a role play game simulation. Hence, we intend to investigate how the usage of social game could increase the level of participation in a collective learning process. This paper presents a general framework encompassing some factors facilitating or hampering the learning capability while designing a participatory role play model process. In order to delineate this model, we have conducted an experiment in two different Norwegian organizations where we asked the participants to define them themselves their learning outcomes while designing a role play. The next section describes the context of study and the adopted methodology in order to conduct our experiments.

3 Context of Study

In order to investigate the effectiveness of Users participation in role play game setting, two Norwegian organizations were selected. The first one is The Norwegian Army Military Academy (NAMA). This institution was keen to participate to the experiment as they wanted to investigate new alternative ways of teaching. The second organization, Tretorget, is an association for small and medium enterprises related to forestry and wood industry sector. This organization aims at providing mechanisms contributing to foster innovation within the member enterprises. Tretorget intends to investigate the best alternative learning methods for their members.

3.1 The Norwegian Army Military Academy

The Norwegian Army Military Academy (NAMA) was established in 1750 and is considered to be one of the oldest active military academies in the world. The school focuses on providing vocational education and training to Norwegian officers that might be sent to war zones. Games have been traditionally used by military schools in order to improve strategies and planning skill of the cadets. The simulation Role Play Game (RPG) techniques are being used today in a variety of military programs designed to enhance the skills of learners. Hence, RPG enables learners to practice necessary skills such as strategy competence in an environment that allows for errors and professional growth without risk. Following this trend of using game, the Norwegian Army Academy has already promoted the use of game in the educational program. Live training might require stringent planning; resources, expensive equipments; therefore virtual simulation or role play simulation can allow some of the activities to be first trained in a safe environment. This approach will not only reduce the cost of the training but will eliminate most the danger associated with real weapons or bad decision making. This is not of course like real action, but this approach is allowed to repeat some actions or learn to improve their decision making process or to understand the behavioral attitude of people while being in experiencing different situations.

This school was selected due to the natural use of games in education. Therefore, we believe that we could get interesting insights in understanding the factors facilitating or hampering the participatory process while designing the rules and defining collectively the learning objectives for a role play game. There were 11 last year cadets volunteering for experiential and experimental learning.

3.2 Tretorget

Tretorget aims to enhance innovation capability in the forestry and wood industry by leveraging competence and skills of people involved in this sector. The objectives are to provide technical and organizational support for fostering entrepreneurship or helping companies to adopt processes leading to innovation. Network building among several stakeholders is seen as essential to build knowledge and to provide means to facilitate the sharing or the transfer of the critical knowledge. Therefore, the organization is keen to investigate new innovative learning approaches since most of the learners come from a low tech industry with in some case very little level of instruction. Conventional training has proved to be not very efficient since participants were not favorable to passive learning.

This organization was selected since the leaders expressed a positive attitude toward exploration of new learning techniques for people having strong practical competences and experiences. We believed that the participatory learning process in order to acquire different type of knowledge should be further investigated. It was interesting to analyze how the active participation of the learners could increase the reflection process which in turn will contribute in the effectiveness of the learning process. In addition, it was important to observe the attitude of the learners while participating in defining the learning objectives and the role play simulation script. After several discussions, the 15 participants expressed their wish to learn more about how to deal with environmental issues that may arise during a working setting.

3.3 Research Methodology

The project was broken down in several phases as follow:

- Definition of the goal of our experiment and methodology delineation in order to conduct our experiments in both organizations
- Raise awareness among the leaders of both organizations through several meetings
- Organization of an interactive workshop with each organization. During this session a group of participants (11 from Nama and 15 from tretorget) had the mission through participatory approach to design collectively a role play game with role and scrip delineation. The participants had to come up with some suggested learning outcomes. They were asked based on previous lectures, experiences,

current issues in the organizations, to come up with some learning outcomes. The idea was to involve participants in the very first step in the topic they feel they need to learn about. The session was intended to understand how people interact and collaborate in designing a narrative game. The focus was voluntary limited to Human resources issues such for i.e conflict solving, communication improvement and so forth. Once the learning objectives have been defined by the both groups, the scenario was played. In this session, the participants shared roles in order to cooperate and learn from each other whilst playing. This also supposed to serve a purpose of impersonalize the input and yet collectively create a role character. The role play session was intended to depict a possible outcome of the script.

- A second session took place after the role play simulation. In this session, we collected data about their individual and collective experience of the first session. This session was based on After Action Review model. Informal interviews have been as well conducted during the session ad the break.
- Based on the literature review, data collected during the both sessions, we have delineated a general model encompassing several factors facilitating or hampering the participatory process while designing the rules and defining collectively the learning objectives for a role play game.
- Validation of the model. The elaborated model needs to be further validated in another type of setting such different organizations or learning outcomes. The project is still ongoing and we are planning to conduct the evaluation process in a near future. We are in the process of looking others organizations willing to test our model.

The experiment was conducted at the within the premises of the NAMA while the workshop for Tretorget took place in a conference room of a hotel. In both cases, the workshop duration lasts for three hours including the break between 2 sessions.

3.4 Data Collection Process

We have conducted our data collection process by adopting a qualitative approach such as observations, focused group interviews and informal discussions during the breaks that has taken place in the interactive workshop. The after action review gave us the possibility to record and analyze some of the attitudes expressed by the participants during the first session while defining the learning objectives of the session. During the first part of the workshop, the participants were defining collectively the learning objectives and co-designing the role play game. Tretorget's participants after a brief discussion, decided quickly and in a smooth manner to learn about how to deal with environmental issues that may arise during a working setting. The consensus was easily reached while deciding to focus on Health Security Environment issues. The Cadets at NAMA were less determined and had a very vague idea of what they wanted to learn. Finally, after lot of discussions and explanation, they decided to define the learning objectives within the field of military tactics.

The researchers and the instructors of both organizations have observed the working collaboration and communication models processes. Attitude and behaviors of participants have been as well observed. An After Action (AAR) review was performed after the first session. This assessment allowed all the participant and the researchers to discover (learn) what happened and why. The goal was to conduct a soft discussion of the previous event in order to learn from that experience. It gave as well the opportunity to analyze and to reflect on the behavioral attitudes and participatory models of the learners.

In addition, it was important to grasp some insights about how the learning process is taking place, how the information should be presented to the learners, how we can use learners experiences and expectations. Additional information is as well important such as how to encourage learners to share and to participate actively in the role play or the collective or individual decision making process is taken.

Prior to the general discussion, a questionnaire investigating their perception on their experience of the role play design was handed out to the participants. The aim of this questionnaire was to trigger a reflection process. The idea was as well to gather data at individual level and then at the group level during the general discussion. The questions were a mix between answers on the Likert scale (nominal) and open ended. Example of questions to the participants include for example, if they felt that they have enough competences to define themselves their own learning objectives, the usability in using RPG as a learning tool, if the benefits or the usability were clearly drawn and so forth. We observed as well as well how the communication and collaboration took place.

The empirical data was analyzed through the use of various tools and methods.

4   Data Analysis

Preliminary observations of the process in both organizations and analysis of the answers of the questionnaire indicate that the participants were very confused and unsecure about what the researchers expected from them. One reason of this general confusion was due to this new approach to trigger a learning process. Being active participants in defining collectively the learning objectives was at first very disturbing. This confusion to some extend was greater at NAMA organization than at Tretorget. Most of the cadets were disoriented by the lack of description of the task to be performed. This is probably due to the fact that the military school usually does not encourage too much participatory process in elaborating learning objectives. It is of course, very important that cadets obey to orders and to be passive receptacles of information. Therefore, they find difficult to break out of the traditional learning model. This is as well stated in the work of Torrs and Inderbitzen that students have difficulties imagining alternative forms of learning and thus require more innovative and transformative learning experiences as central components of today's liberal education [27].

The general attitude was a negative feeling of this new approach to learn something that they have defined themselves. After some discussions, they decided to define the goal of the learning tactic in a military context. During the discussion, it was interesting to observe the communication patterns of the participants. The script/scenario delineation engaged somewhat a lot of discussion as they could not see the benefits of such approach. It was a bit surprising as we expected them to have already some experiences in using simulation tool in their education. But the overall negative feeling about the whole experiment led them to have a lot of discussions highlighting the little pertinence of this participatory process in the design of the role play. This was probably as well due to the selected topic. Military tactic is a wide domain and the lack of clarity in defining a more focused learning objective might have influence their perception of the difficulty of writing a scrip or playing a role in the simulated situation.

The learners at NAMA that do report on the role playing having an impact on the learning, respond that it gave them new insight on their plan. However, the instructor was positive towards the RPS and could in the excerpt from the written dialogue point to increased understanding of the tactics curriculum.

The learners at Tretorget have a profile of professional with a very pragmatic and practical working life experience. Although they expressed some surprises on the required task for the session, they quickly came up with the idea to define a role play on the drug or alcohol abuse at the workplace. They actually found out that being involved at earlier stages of the role play design was a good approach as they get the possibility to adopt a consensual approach to define collectively the learning goal. In addition, people tend to construct knowledge in a best way when they are engaged personally, therefore the content-based scenario proved to be interesting alternative teaching approach. Furthermore, active participation in tasks has demonstrated that role-play games could be considered as a very good means to encourage every player (learner) to take individual responsibilities, practices real-life simulated tasks and experience the outcomes of any decision making.

The learners at Tretorget also reported on gaining a new insight, not only about the collective process of coming up with a scenario, but also with regards to the topic they had chosen. Some of what they had learned during the lecturing on the course was confirmed and even if this was no surprise, it was something different to "experience" it.

Based on data analysis, we have delineated a model encompassing factors that contribute to the efficiency of a participatory learning process (Fig. 2). Individual and organizational factors, concrete experience and reflective learning catalyze an effective participatory learning. The four components that contribute to enhance the participatory learning effectiveness are user motivation, Organizational involvement, concrete experiences and the reflective learning outcomes.

In order to foster user motivation, the participants claimed during second session of the workshop for both organizations, that they had to see clear potential benefits, or the usability of such method, adequate skills was needed. Rewards or incentives were as well important for increasing the participation in the learning process. Participants suggested that getting enough support and sufficient time to use the RPG

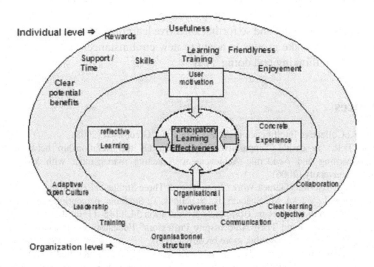

Fig. 2 Participatory learning effectiveness model

as a learning tool was vital. They also felt that the script/scenario design should provide a high level of enjoyment and it was only possible if the people awareness was sufficient through training.

At organizational level, it is important to specify a clear communication flow, or provide some adequate mechanisms supporting an effective training. The top management involvement is considered as a key to introduce new innovative teaching or learning methods. It was important for the group to be able to cooperate and to be open to a new change. Therefore, we believe that by taking into account some of the individual /organizational factors and providing mechanisms to reflect on the learning, the participatory learning should be more effective.

5   Conclusion

This paper discusses the concepts in using role play game simulation in order to foster user's participation in a collective learning effort. In order to validate our assumptions or hypothesis, we have conducted two experiments involving two Norwegian organizations. The goal was to investigate the evidences of using role play game in a participatory as a suitable tool fostering collective learning in a group of people. To this end, we have conducted an interactive workshop encompassing two sessions.

Our investigation has demonstrated that Role play game is a powerful tool to foster collective learning capability by creating group dynamics, team's development.

Based on the collected data, we have outlined a general model encompassing factors facilitating the participatory learning in a role play context such as communication patterns, decision making and so forth. Collective learning and developing collective competencies can make groups to adapt to new circumstances and to develop new competencies in thinking and doing. [19]

References

1. Alag, S.: Collective Intelligence in Action. Manning, Greenwich (2008)
2. Blunt, D.R.: A casual-Comparative Exploration Of the Relationship between Game-based learning and Academic Achievement: Teaching management with Video games, Walden University (2006)
3. Blunt, R.: Do Serious Games Work? Results from Three Studies. eLearn magazine (2009)
4. Bourgonjon, J., Valcke, M., Soetaert, R., Schellens, T.: Students' perceptions about the use of video games in the classroom. Computers & Education 54, 1145–1156 (2010)
5. Carlson, S.: Can Grand Theft Auto Inspire Professors? Educators say the virtual worlds of video games help students think more broadly (2003),
   http://chronicle.com/free/v49/i49/49a03101.htm (cited 2007)
6. Chesler, M., Fox, R.: Role-playing methods in classroom. Science Research Associates, Chicago (1966)
7. Dewey, J.: Experience & Education. Touchstone, New York (1938)
8. Dewey, J.: Democracy and Education. Barnes & Noble Books (2005)
9. Dionnet, M., Kuper, M., Hammani, A., Garin, P.: Combining role-playing games and policy simulation exercises: An experience with Moroccan smallholder farmers. Simulation and Gaming 39, 498–514 (2008)
10. Eikeland, O., Berg, A.M.: Medvirkningsbasert organisasjonslæring og utviklingsarbeid i kommunene Kommuneforlaget, Oslo (1997)
11. Gee, J.P.: What video games have to teach us about learning and literacy. Macmillan (2003)
12. Kilgore, D.W.: Understanding learning in social movements: a theory of collective learning. International Journal of Lifelong Education 18, 192–202 (1999)
13. Knowles, M.S.: The adult learner: a neglected species. Gulf Pub. Co., Houston (1990)
14. Kolb, D.A.: Experiential learning: experience as the source of learning and development. Prentice-Hall, Englewood Cliffs (1984)
15. Lazaric, N.: The role of routines, rules and habits in collective learning: Some epistemological and ontological considerations. European Journal of Economic and Social Systems 14, 157–171 (2000)
16. Lee, K.E., Lee, J.V.: Benefits of a virtual learning environment in enabling collaborative and constructivist learning in a community of specialist nursing practice. In: Proceedings of the IEEE International Conference on Advanced Learning Technologies (2004)
17. Loui, M.C.: What Can Students Learn in an Extended Role-Play Simulation on Technology and Society? Bulletin of Science Technology Society 29, 37–47 (2009)
18. Michael, D., Chen, S.: Serious Games: Games That Educate, Train and Inform. Thomson Course Technology, Boston (2006)
19. Pivec, M., Dziabenko, O.: Game-Based Learning in Universities and Lifelong Learning: UniGame: Social Skills and Knowledge Training. Game Concept. Journal of Universal Computer Science 10, 4–16 (2004)
20. Quinn, C.N.: Engaging Learning: Designing e-Learning Simulation Games. Pfeiffer, San Francisco (2005)
21. Reynolds, M., Vince, R. (eds.): The Handbook of Experiential Learning & Management Education. Pfeiffer, Oxford (2007)
22. Rogers, C.R.: Freedom to learn. Merrill, Colombus (1969)

23. Schnotz, W.: Towards an integrated view of learning from text and visual displays. Educational Psychology Review 14, 101–120 (2002)
24. Shih, J.-L., Shih, B.-J., Chen, R.-L.: The design and evaluation of virtual situation Role-playing game VSRPG. International Journal of Instructional Technology and Distance Learning 3 (2006)
25. Silberman, M. (ed.): The Handbook of Experiential Learning. Pfeiffer, San Francisco (2007)
26. Tippelt, R., Amorós, A.: Innovative and participative learning-teaching approaches within a project based training framework (2003),
    http://star-www.inwent.org/starweb/inwent/docs/
    Lehrbrief_03_engl.pdf (cited January 12, 2011)
27. Torrs, D., Inderbitzin, M.: Imagining a Liberal Education Critically Examining the Learning Process Through Simulation. Journal of Transformative Education 4, 175–189 (2006)
28. Wenger, E.: Communities of practice. Cambridge University Press, Cambridge (1998)
29. Young, L.: Bridging Theory and Practice: Developing Guidelines to Facilitate the Design of Computer-based Learning Environments. Canadian Journal of Learning and Technology 29 (2003)

A Complex Network Analysis of the Weighted Graph of the Web2.0 Service Network

Kibae Kim and Jörn Altmann

Abstract. Service providers that own Web2.0 services allow Internet users not only to access their Web2.0 services but also to create new Web2.0 services (mashups) based on theirs. This creation of mashups generates the Web2.0 service network, in which a node represents a Web2.0 service and a link between two nodes represents a mashup using the two Web2.0 services linked. Since this Web2.0 service network is constructed without the control of a single entity (i.e., it is self-organizing), the network topology of the Web2.0 service network shows the scale-free characteristic. With respect of the weighting of those links, however, there are different approaches. Prior research either considered binary links or links that are weighted by summing up the number of mashups. Since the last approach might overestimate the strength of the link, we calculate the link weights according to Newman's approach in this paper. Based on this weighted graph of the Web2.0 service network, we investigate the topology of the weighted graph and examine the pattern of Web2.0 service creations. Our results show that the Newman-based weighted graph of the Web2.0 service network shows the characteristics of a scale-free network and a small-world network.

Keywords: Web2.0 services, mashup, network topology analysis, self-organized networks, small-world networks, scale-free networks, complex networks.

Kibae Kim
Institut für Datentechnik und Kommunikationsnetze Technische Universität Braunschweig
Hans-Sommer-Str. 66, 38106 Braunschweig, Germany
e-mail: kim@ida.ing.tu-bs.de

Jörn Altmann
Technology Management, Economics, and Policy Program & Department of Industrial Engineering, College of Engineering Seoul National University 599 Gwanak-Ro, Gwanak-Gu, Seoul 151-742, Korea
e-mail: jorn.altmann@acm.org

J. Altmann et al. (Eds.): Advances in Collective Intelligence 2011, AISC 113, pp. 79–90.
springerlink.com   © Springer-Verlag Berlin Heidelberg 2012

1 Introduction

Many Web2.0 service providers allow free access to their Web2.0 services through open Application Programming Interfaces (API). An Internet user (in the role of a Web2.0 service developer), who is able to apply Web2.0 technologies (e.g., Asynchronous JavaScript and XML), can combine these Web2.0 services to develop new ones [14, 20]. These new Web2.0 services, which are called Web2.0 mashups, or simply mashups [7, 26, 27], might also include the Internet user's own data and functions.

To analyze the creation of Web2.0 services, we introduce the Web2.0 service network. A node of the Web2.0 service network represents an existing Web2.0 service and a link between two nodes represents the existence of a mashup using these two Web2.0 services.

Since the Web2.0 service network is based on user decisions, we expect the Web2.0 service network to be self-organized. In particular, we expect the Web2.0 service network to show features that appear in complex social networks (e.g., random networks, small world networks, and scale-free networks).

For about a decade, prior research has examined mechanisms for generating network topologies that are similar to those of self-organized social networks. Watts and Strogatz showed that the random rewiring model can transform a regular network into a small-world network by randomly rewiring links of the regular network [24]. In small world networks generated by Watts and Strogatz, a node is embedded in a small, cohesive cluster that can reach any other node in the network through a very few intermediaries only.

Furthermore, empirical studies of the WWW, open source social networks, and academic collaboration networks suggest that a majority of self-organized networks are scale-free. Under this property, the degree frequency decays following a power function [1, 6, 17, 22, 23]. The nodes that are located in the long tail of the power function are called hubs. They make the network diameter small [4].

It is important to analyze the characteristics of these self-organized networks, since the characteristics affect the performance of communication within the network and the robustness of the network.

With respect to information diffusion in networks, the average shortest path is important. Information diffuses quickly, if the information initiates at a hub within a scale-free network [12, 13].

Links within a scale-free network are concentrated on a few hubs while links are equally distributed among all nodes in small world and random networks. Therefore, scale-free networks are not robust to intentional attacks comparing to random networks [2]. On the other hand, due to the same reason, scale-free networks are stronger to random failures than random networks.

Several studies examined the topology of networks related to Web2.0 services. Fu et al. analyzed the network topology of social networking services. Hwang et al. examined the Web2.0 service network to show its scale-free property [8, 9]. Kim et al. investigated the openness between subgroups, showing that openness impacts the evolution of complex network and, in particular, Web2.0 service networks [11].

However, these studies do not consider the weights of the Web2.0 service network comprehensively. Although Hwang et al. (2009) and Kim et al. (2010) considered weights in the Web2.0 service network, the calculation of a weight of a link is the simple summation of the number of mashups [9, 11]. This calculation of weights may overestimate the strength of a link between two nodes. For example, consider a mashup i that is created with two Web2.0 services (i.e., x and y) and a mashup j that is created with three Web2.0 services (i.e., x, y, and z). The contribution of each of the two Web2.0 services x and y to mashup i is larger than the contribution of each of the three Web2.0 services x, y, and z to mashup j. Therefore, the link through mashup j should be weighted less than the link through mashup i. In order to account for this, we follow Newman's approach. Newman calculated the weight of a link in an academic collaboration network as the sum of all collaboration divided by the total number of co-authors minus one [17], avoiding overestimation.

In this paper, we investigate the topology of the Web2.0 service network that has been established according to Newman's approach [17]. In our analysis, we check whether the pattern of the mashup creation corresponds to that of well-known self-organized networks. In particular, we examine whether the Web2.0 service network shows small-world network and scale-free network properties. Finally, we compare the result with previous research on Web2.0 service network.

The remainder of this paper is organized in 3 sections. The next section introduces the literature on node degree distributions, small-world networks, scale-free networks, and on weighted graphs. In section 3, the Web2.0 service network is analyzed with respect to indicators of small-world networks and scale-free networks. In section 4, we conclude the paper with a discussion on the mashup creation within the Web2.0 service network.

2 Theoretical Background on Complex Networks

2.1 Classification of Node Degree Distributions

The topologies of networks can be classified into two groups according to the type of heterogeneity of node degree distributions. The first group comprises networks, which node degree distribution varies around a mean value (i.e., homogeneous node degree distribution). Examples of those networks are regular networks, small-world networks, and random networks (e.g., Poison distributed random network) [24]. These networks have no node with a distinctive central position. All nodes have a node degree that is in a range around the mean value of the node degree in the network. The only difference between these three networks is that the topology of a small-world network shows a higher variance around the mean value (i.e., more randomness) than the topology of a regular network, and the topology of a random network shows a higher variance around the mean value (i.e., more randomness) than the topology of a small-world network.

The second group of networks is identified through a node degree distribution that does not vary around the node degree mean value. These networks are identified through the existence of distinctive central nodes, which are connected to many nodes, through many nodes with only one neighbor, and through the lack of nodes with more than one common neighbor (i.e., heterogeneous node degree distribution). Examples of this group are scale-free networks [2, 3] and tree-structured networks. A tree network has a heterogeneous node degree distribution, since a central node stretches to terminal nodes through a few intermediary nodes. A star network is a tree network where all nodes have exactly one link except for one node that connects all other nodes.

2.2 Small-World Networks

Watts and Strogatz generated a small-world network by rewiring links of a regular network with probability p [24]. They used the average clustering coefficient to measure how the network is clustered, and the average shortest path length to evaluate how near the members in the network are.

The clustering coefficient $Cl_v(g)$ of node v in graph g is defined as the ratio of the number of links between neighbors of node v, to the theoretically maximum number of links between them:

$$Cl_v(g) = \frac{\sum_{i,j \in N_v(g)} a_{ij}}{k_v(k_v - 1)/2} \tag{1}$$

where a_{ij} is an element of adjacency matrix A. a_{ij} is 1 if node i is connected with node j, otherwise 0. $N_v(g)$ denotes the set of neighbors of node v. k_v represents the degree of node v (i.e., the number of links of node v), which is calculated as

$$k_i = \sum_j a_{ij} \tag{2}$$

The average clustering coefficient is calculated as the average of the clustering coefficient of all nodes v.

The shortest path length $d(i, j)$ between node i and node j is defined as the number of links in the shortest path. The average shortest path is defined as the average of the shortest paths between any two nodes i and j.

Regular networks have the same number of links per node. Consequently, the degree distribution shows a peak at a certain degree. In this kind of network, each node is embedded in a cluster (i.e., high average clustering coefficient) but is far from other nodes (i.e., long average shortest path length). If the rewiring probability (i.e., the probability that a link is chosen to be rewired) goes to 1 (i.e., all links in the regular network are rearranged), the network transforms to a Poisson distributed random network. In this random network, nodes rarely share neighbors (i.e., low average clustering coefficient) and are close to each other (i.e., short average shortest path length).

The small-world network is positioned between the Poisson distributed random network and the regular network. It is highly clustered (i.e., high average clustering coefficient) and has nodes that are connected closely (i.e., short average shortest path length) [24]. It implies that the nodes in the small-world network are connected to each other efficiently. It was empirically tested by Milgram that a real world social network has a small world-network topology [15]. The small-world topology implies that a person frequently meets a friend of his friends.

2.3 Scale-Free Networks

Empirical studies of the WWW, open source social networks, and academic collaboration networks suggest that a substantial number of self-organized networks are scale-free [1, 6, 17, 22, 23]. Scale-free networks are networks, whose frequency $P(k)$ of node degree k decays by a power function k^γ. Prior research explained that a scale-free network is generated by a simple rule such as the preferential attachment rule or the rule of merging and regeneration [3, 10, 21].

Generally, noise appears in the high node degree range, which represents important information about a scale-free network, due to the low frequency in this range. In order to eliminate the problem of low frequency, the cumulative degree distribution is more useful. It determines the scale-free characteristic of a network by calculating $k_>$, i.e., the frequency of node degrees larger than k [18]:

$$P(k_>) = \int_k^\infty P(k)dk \sim k^{-(\gamma-1)} \tag{3}$$

The distribution of a scale-free network is heterogeneous, compared to the distributions of Poisson distributed random networks and small world networks. It implies that a few hubs have a large number of links and a large part of nodes has only a small number of links [1, 4].

2.4 Weighted Graphs

So far, existing literature showed only the previously mentioned properties of binary networks. In those binary networks, the element a_{ij} of an adjacency matrix A indicates only the presence of a link between node i and j (i.e., the neighborhood of a node). Although the weight of a link also exhibits important information about networks, it has not been considered until now. In particular, the weights show how frequently the nodes interact with each other.

The weights can be defined in different ways, mainly depending on the context of the network analysis. For example, the weight of a link between two adjacent routers in the Internet can be defined as the data traffic capacity between them. The degree of intimacy, or goodwill, between two friends is also a good example for expressing it with a link weight.

Newman suggested a measure for capturing the strength of collaboration between two coauthors of an academic article [16]. He assumed that the strength of collaboration between two authors is proportional to the number of articles collaborated on, and that it is inversely proportional to the number of authors participating in the article. Under this assumption the weight w_{ij} of a link between author i and author j in an academic collaboration network is defined as:

$$w_{ij} = \sum_p \frac{\delta_i^p \delta_j^p}{n_p - 1} \tag{4}$$

δ_i^p represents the contribution of author i to paper p. δ_i^p is 1 if author i contributed to paper p, otherwise 0. n_p is the number of authors who contributed to paper p.

With the weights between node i and node j defined, the complex network indicators for weighted graphs can be defined corresponding to the indicators for binary networks. The strength of the degree of node i in weighted networks is defined (analogous to the node degree in a binary graph) as the sum of link weights that node i involves [5, 16, 19]:

$$s_i = \sum_j w_{ij} \tag{5}$$

Furthermore, the definitions of path length and clustering coefficient need to be adapted. For the weighted shortest path length, $d^w(i, j)$ between node i and node j is defined as the minimum sum of inversed weights of the links on the path between node i and node j [19]:

$$d^w(i, j) = \min \left(\frac{1}{w_{ih}} + \ldots + \frac{1}{w_{gj}} \right) \tag{6}$$

If each weight goes to 1, or if each weight implies only the presence of a link, the equation of the weighted shortest path length $d^w(i, j)$ is reduced to that of a binary graph.

With respect to the clustering coefficient, it is assumed that a node is clustered more cohesively with its neighbors if its link weights are greater than 1. The definition of the clustering coefficient Cl_v of node v has been extended by Barrat et al. (2004) to capture this assumption [5]. We corrected their definition by multiplying it with a factor 2, in order to get $0 \leq Cl_v^w \leq 1$:

$$Cl_v^w = \frac{2}{s_v(k_v - 1)} \sum_{i,j} \frac{w_{vi} + w_{vj}}{2} a_{vi} a_{vj} a_{ij} \tag{7}$$

Here, each triplet $(a_{vi} a_{vj} a_{ij})$ is weighted with the average link weight obtained from the two links from node v to the two neighboring nodes i and j. This sum of the weighted triplets is normalized by $s_v(k_v - 1)/2$, considering the strength of the degree of node v.

3 Analysis

3.1 Description of Data

A list of Web2.0 services and mashups has been gathered from the Website www.programmableweb.com. It includes 445 Web2.0 services with open APIs and 1,929 Web2.0 mashups. These services have registered with the Website between September 1, 2005 and May 21, 2007. From this initial data set, 214 Web2.0 services were deleted in the data set, since they were not used to develop any Web2.0 mashup during the study period. Furthermore, we eliminated 15 Web2.0 services since they were solely used for mashup creation (i.e., each of these Web2.0 services was not used together with any other Web2.0 service for creating a mashup. The deletion of these Web2.0 services does not affect the network characteristic, since these Web2.0 services are isolated in the Web2.0 service network. Consequently, the number of Web2.0 actually used for the analysis is 216.

The Web2.0 service network was constructed as described earlier. A node of the Web2.0 service network represents an existing Web2.0 service and a link between two nodes indicates which two Web2.0 services were jointly used to create a new Web2.0 mashup.

Therefore, an element a_{ij} of the adjacency matrix A of the Web2.0 service network is 1, if the two Web2.0 services were used at least once to create a Web2.0 mashup. Otherwise, a_{ij} is equal to 0. Within the generated Web2.0 service network, 1929 Web2.0 mashups generated 1916 links. The average degree of a node is $k = 8.870$.

The weights of the Web2.0 service network were calculated following Newman's approach [17]. In equation (4), it was assumed that δ_i^p is 1, if Web2.0 service i was used to create mashup p, otherwise 0. As a result, the total weight calculated is 924.738 for the 1916 links in the Web2.0 service network.

3.2 Properties of Weights

The distribution of the link weights of the Web2.0 service network shows a power law distribution. The logarithm of the cumulative number of weights decays linearly according to the logarithm of the weight (Fig. 1a). The result of the linear regression on this data shows that the slope is 1.945, the significance probability is 0.000 and the determination coefficient R^2 is 0.974. That means, the distribution function of the weight w is $P(w) \sim w^{-2.945}$, and the relationship is statistically significant.

The strength s of a node tends to increase exponentially as the node degree k of the node increases. The logarithm of the strength $s(k)$ conditional on k shows a slightly positive linear correlation with k (Fig. 1b). The slope in the log-linear regression model is 0.013. The significance probability is 0.000 and the determination coefficient R^2 is equal to 0.450. These results exhibit that Web2.0 services, which have been combined with many different Web2.0 services, are utilized more frequently to develop mashups.

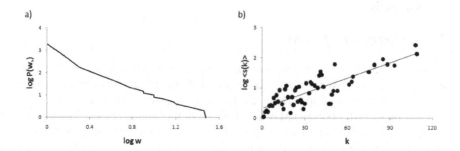

Fig. 1 Properties of the weights of the Web2.0 service network: a) the weight distribution in the log-log scale; and b) the relationship between the node degree k and the logarithm of the strength of the node degree

3.3 Small-World Properties

To evaluate the small-world property of the weighted graph of the Web2.0 service network, we measure the weighted clustering coefficients and the average weighted shortest path length of the network. The average weighted clustering coefficient of the Web2.0 service network is calculated as 0.594 The standard deviation of the weighted clustering coefficient is calculated as 0.476. The weighted clustering coefficients of the Web2.0 service network range from 0 (theoretical minimum) to 1 (theoretical maximum). Among 216 nodes, 68 nodes have a clustering coefficient of 1, and 35 nodes have a coefficient of 0.

The calculation of the average weighted shortest path length of the Web2.0 service network results in 5.99. The standard deviation of the average weighted shortest path length is 13.443. Because of the large standard deviation, it is hard to say that the average weighted path length characterizes the distance between any nodes.

In order to interpret the measured small-world indicator values (i.e., weighted clustering coefficient and the average weighted shortest path length) of the Web2.0 service network, we compare them with indicator values of artificially constructed small-world networks that have the same number of nodes, the same average node degree, and the same link weight distribution. The artificially small-world network, which we consider, is created using the random redirecting process of Watts and Strogatz [24]. The random redirection process starts with a regular network that has 216 nodes. Each node of this regular network has 8 neighbors.

The link weights are allocated to the links of the network randomly, following the power distribution function $P(w) \sim w^{-2.945}$ where w indicates the weight. This is based on the method of Yang and Yu [25].

The average weighted clustering coefficient $Cl^w(0)$ and the average weighted shortest path length $L^w(0)$ of the regular network (i.e., node degree k of each node is 8 and the rewiring probability p equals 0) are 0.643 and 14.102, respectively. The values of $Cl^w(0)$ and $L^w(0)$ are used for normalizing the values obtained for networks that are derived using different rewiring probabilities p.

As the probability p that a link is redirected increases, the two normalized indicator values decrease (Fig. 2a). While the normalized average weighted shortest path length decreases quickly with small p, the average weighted clustering coefficient keeps a high value until $p \approx 0.10$ before it decreases rapidly as well. Fig. 2a shows that there is a range between $p = 0.01$ and $p = 0.1$ with a high normalized average weighted clustering coefficient (i.e., larger than about 0.8) and a low normalized average weighted shortest path length (i.e., smaller than about 0.6), indicating small-world networks.

The normalized average weighted clustering coefficient of the Web2.0 service network normalized by that of a regular network with a node degree of 8 is 0.924. The normalized average weighted shortest path length is 0.425. The comparison of the two small-world indicators of the Web2.0 service network with the values shown in Fig. 2a suggests that the Web2.0 service network is located in the small-world area (Fig. 2b). In Fig. 2b, the area at the right lower corner is the small-world area that has been defined by Watts and Strogatz vaguely [24]. In addition to this, the Web2.0 service network marks a point almost adjacent to a small-world network. It has to be noted that the curve of small-area networks (Fig.2b) shows a kink in the small-world network area. The location and strength of the kink needs to be analyzed in future research though.

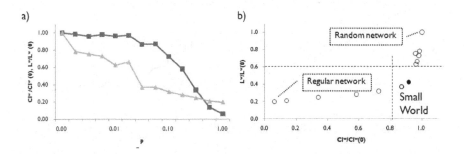

Fig. 2 Small-world property of the Web2.0 service network: a) the normalized average clustering coefficient and the normalized average weighted shortest path length of a network generated by the random rewiring mechanism; and b) the relationship between the normalized average weighted shortest path length and the normalized average clustering coefficient for networks generated using the random rewiring process (circle), and for the Web2.0 service network (filled black dot).

3.4 Scale-Free Properties

To examine the scale-free property of the weighted graph of the Web2.0 service network, we use the strengths of node degrees. The cumulative strength distribution of the Web2.0 service network looks like an ordinary scale-free network. The cumulative frequency $P(s_>)$ of strength s decays linearly in the log-log scale coordinate plane (Fig. 3a). The log-log regression of the distribution results in a slope of -0.953, a significance probability of 0.000, and a determination coefficient R^2 of 0.978. That means, it is statistically significant that the weighted graph of the Web2.0 service network has the strength distribution of the power law:

$$P(s) \sim s^{-1.953} \tag{8}$$

In order to avoid mistaking the exponential decaying for the power law, we draw the cumulative strength distribution in a log-linear scale coordinate plane (Fig. 3b). With a distribution $P(k) \sim \exp(-\zeta k)$, the cumulative distribution $P(k_>)$ can be calculated to be:

$$P(s_>) = \int_s^\infty P(s')ds' = \int_s^\infty \exp(-\zeta s')ds' = \exp(-\zeta s) \tag{9}$$

If the distribution of the node degree strength is exponentially decaying, the cumulative strength distribution should be linear in the log-linear scale coordinate plane. However, since the cumulative strength distribution in the log-linear scale is convex (Fig. 3b), we conclude that the weighted graph of the Web2.0 service network shows a scale-free property.

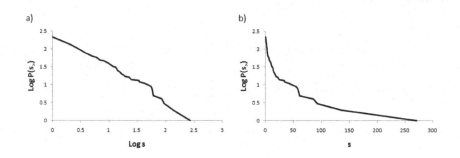

Fig. 3 The cumulative degree distribution of the Web2.0 service network in a) the log-log scale coordinates plane and in b) the log-linear scale coordinates plane.

4 Discussion and Conclusion

To analyze the behavioral pattern of the mashup creation on the Web2.0 service network, we measured the average weighted clustering coefficient, the average weighted shortest path length, and the node degree strength of the Web2.0 service network. Based on the analysis results, two conclusions can be drawn: First, the Web2.0 service is a small network as specified by Watts and Strogatz [24]. While the average weighted clustering coefficient of the network is close to the maximum possible value, the average weighted shortest path length is very small in the Web2.0 service network.

Second, the Web2.0 service network has a node degree strength distribution, which looks like a scale-free network. The cumulative strength distribution of the Web2.0 service network significantly fits well into a power function. These results suggest that the weighted Web2.0 service network is an ordinary scale-free network and a small-world network. This result is interesting because it is contrary to the result of the prior research of the Web2.0 service network [9]. Hwang et al. (2009) suggested that the Web2.0 service network has an extraordinarily low exponent of the power law distribution. The difference of our analysis from Hwang et al. (2009) is only how the link weights are calculated. We used Newman's approach for the weight calculation [16, 17], while Hwang et al. (2009) summed up all creations of mashups for each pair of nodes. This dependency of the network topology on the weighting technique requires further investigation of the Web2.0 service network.

Although we analyzed the pattern of mashup creation in the Web2.0 service network comprehensively, this research involves two limitations. First, self-links within the Web2.0 service network were omitted in this research. Although a new mashup can be generated in two ways, namely by adding value to an existing Web2.0 service (self-link) or by combining several Web2.0 services, we considered the latter one only. Second, we randomly weighted links in the generated complex networks to compare them with the Web2.0 service network. However, we have checked that the node degree strength is correlated with the node degree. To conduct a more proper comparison, we need a weighting scheme which weights links depending on the degrees of two nodes at its end.

References

1. Albert, R., Jeong, H., Barabási, A.-L.: Diameter of the World-Wide Web. Nature 401, 130–131 (1999)
2. Albert, R., Jeong, H., Barabási, A.-L.: Error and Attack Tolerance of Complex Networks. Nature 406, 378–382 (2000)
3. Barabási, A.-L., Albert, R.: Emergence of Scaling in Random Networks. Science 286, 509–512 (1999)
4. Barabási, A.-L.: Linked: The New Science of Networks. Perseus Pub., Massachusetts (2002)
5. Barrat, A., Barthélemy, M., Pastor-Satorras, R., Vespignani, A.: he Architecture of Complex Weighted Networks. P. Natl. Acad. Sci. USA 101, 3747–3752 (2004)
6. Björneborn, L.: 'Mini small worlds' of shortest link paths crossing domain boundaries in an academic space. Scientometrics 68, 395–414 (2006)
7. Feiler, J.: How to do Everything with Web2. McGraw-Hill, New York (2008)
8. Fu, F., Liu, L., Wang, L.: Empirical Analysis of Online Social Networks in the Age of Web2.0. Physica A 387, 675–684 (2008)
9. Hwang, J., Altmann, J., Kim, K.: The Structural Evolution of the Web2.0 Service Network. Online Inform. Rev. 33, 1040–1057 (2009)
10. Kim, B.J., Trusina, A., Minnhagen, P., Sneppen, K.: Self Organized Scale-Free Networks from Merging and Regeneration. Eur. Phys. J. B 43, 369–372 (2005)

11. Kim, K., Altmann, J., Hwang, J.: Measuring and Analyzing the Openness of the Web2.0 Service Network for Improving the Innovation Capacity of the Web2.0 System through Collective Intelligence. In: 1st Symposium on Collective Intelligence, Hagen, Germany (2010)
12. Kiss, C., Bichler, M.: Identification of Influencers: measuring Influence in Customer Networks. Decis. Support Syst. 46, 233–253 (2008)
13. Kuandykov, L., Sokolov, M.: Impact of social neighborhood on diffusion of innovation S-curve. Decis. Support Syst. 48, 531–535 (2010)
14. Lai, L.S.L., Turban, E.: Groups Formation and Operations in the Web2.0 Environment and Social Networks. Group Decis. Negot. 17, 387–402 (2008)
15. Milgram, S.: The Small-World Problem. Psychol. Today 1, 61–67 (1967)
16. Newman, M.E.J.: Scientific Collaboration Networks: II. Shortest Paths, Weighted Networks and Centrality. Phys. Rev. E 64, 016132 (2001)
17. Newman, M.E.J.: Clustering and Preferential Attachment in Growing Networks. Phys. Rev. E. 64, 025102 (2001)
18. Newman, M.E.J.: Power Laws, Pareto Distributions and Zipf's Law. Contemp. Phys. 46, 323–351 (2005)
19. Opsahl, T., Agneessens, F., Skvoretz, J.: Node Centrality in Weighted Networks: Generalizing Degree and Shortest Paths. Soc. Networks 32, 245–251 (2010)
20. O'Reilly, T.: What is Web2.0: Design Patterns and Business Models for the Next Generation of Software. Commun. Strat. 65, 17–37 (2007)
21. Park, K., Lai, Y.-C., Ye, N.: Self-Organized Scale-Free Networks. Phys. Rev. E 72, 026131 (2005)
22. Valverde, S., Solé, R.V.: Self-Organization versus Hierarchy in Open Source Social Networks. Phys. Rev. E 76, 046118 (2007)
23. Wagner, C.S., Leydesdorff, L.: Network Structure, Self-Organization, and the Growth of international Collaboration in Science. Res. Policy 34, 1608–1618 (2005)
24. Watts, D.J., Strogatz, S.H.: Collective Dynamics of 'Small-World' Networks. Nature 393, 440–442 (1998)
25. Yang, Y., Yu, X.: Weighted Small World Complex Networks: Smart Sliding Mode Control. In: Huang, D.-S., Jo, K.-H., Lee, H.-H., Kang, H.-J., Bevilacqua, V. (eds.) ICIC 2009. LNCS, vol. 5755, pp. 935–944. Springer, Heidelberg (2009)
26. Yee, R.: Pro Web2.0 Mashups: Remixing Data and Web Services. Apress, New York (2008)
27. Zammetti, F.W.: Practical Javascript, DOM Scripting, and Ajax Projects. Springer, New York

How Collective Intelligence Redefines Education

Lynn Ilon

Abstract. While collective intelligence systems become ubiquitous for learning in knowledge industries, civic life and personal lives, they have yet to be embraced into formal schooling systems. Still, learning, knowledge and assessment protocols adhere, in large part, to the educational system's logic of the industrial era. The temptation is to view schooling as falling behind with teacher retraining and curriculum revision leading the way. This article examines the underlying logic of both collective intelligence and formal education systems and traces education's reluctance to it roots in an industrial era and the incentives prevailing in its structures. Embracing collective intelligence, then, will require a redefinition of schooling rather than a mere retooling.

1 Introduction

While the forces of collective intelligence build global learning systems, content and networks, formal schools continue to teach, test and organize themselves much as they did fifty years ago, before technology was available. Although some classrooms allow students to use the internet for selected investigation of topics, the notion that knowledge can be built by groups of people, over a global network, using self-regulated norms is antithetical to the environment of learning within today's formal educational systems.

This seems puzzling given that collective intelligence is behind the large-scale learning of many powerful systems today. Google, Linex, Wikipedia and numerous user help centers are accessed constantly as learning tools throughout the world – in business, civic life, government, social circles, religious communities, home life and personal growth. Some schools even allow their access. But the logic behind them,

Lynn Ilon
Seoul National University
e-mail: lynnilon@snu.ac.kr

J. Altmann et al. (Eds.): Advances in Collective Intelligence 2011, AISC 113, pp. 91–102.
springerlink.com   © Springer-Verlag Berlin Heidelberg 2012

that of collective knowledge, is not explored as a learning tool. Students, teachers and administrators still work for a basis of expertise knowledge.

Most work which investigates school resistance to such environments takes a view that teachers are not equipped with the technical skills and need additional training. Other literature suggests that they are professionally resistant. This article takes a different view. It analyzes the underlying economic and structural forces of formal education and contrast them with that of collective intelligence. In doing so, it demonstrates that the any large-scale incorporation of collective intelligence implies not just a technological change or change in attitude of teachers, but rather a system change in the way that formal education is conceived – a redefinition of education.

The paper begins by describing the major characteristics of formal educational systems. These descriptions reveal that educational system characteristics cannot easily be meshed with collective intelligence usage. This raises the question of why such incompatible characteristics are retained in a time of ubiquitous collective intelligence. The second part of the paper explores this question demonstrating that the underlying logic of the two systems are incompatible and unlikely to merge easily. The third part of the paper outlines the shift in economics that bred an environment for collective intelligence but, at the same time, represents a wholesale redefinition of formal education systems. This redefinition requires more than training and technical adjustments and underlies the resistance to integration of collective knowledge into formal education systems.

2   Education as Managed Intelligence

Formal schooling systems are designed explicitly and philosophically to be controlled from the top. This is in keeping with their notion that expertise comes from the top and that top-down management serves the public good. Thus, collective intelligence cannot be viewed as a viable alternative or even a viable large-scale addition. Its sytems-based, self-regulating and dynamic characteristics are not so much technically difficult to integrate as they are oppositional to the foundational principles of formal schooling.

Formal education systems were designed to give people the knowledge and skills that they could not obtain in their everyday environments. Mass education centrally controlled through government began with the industrial revolution when farm-based and community based education needed to move to a more expert level of reading, math and professional expertise. The values embedded, at this time, were that the teachers, professors and the books they wrote represented expert knowledge that had been vetted. Other knowledge could not be assumed to be correct. One mark of a good scholar was to know the difference between expert knowledge and "common knowledge." Hence, for example, use of Wikipedia is often banned in schools.

The industrial model had another influence. The top-down, functional management system of factories was adopted. Formal educational systems are guided by a set of policies and procedures defined, usually, by central governments and meant to deliver a consistent, known and knowable set of knowledge, values and skills to children. The public employees that guide these massive national systems seek to guarantee the public that its investment in education produces a product (educated adults) that meets generally acceptable standards and does so in an environment that is safe, controlled and efficient even as each teacher works, generally, in isolation for much of the day [4, 20]. In order to deliver these outcomes, particular characteristics are endemic and differ substantially from the fundamental precepts of collective intelligence.

2.1 System Controlled

Educational systems are viewed as bureaucratic systems which must be controlled by government in order to maximize societal benefits [19, 22]. If governments do not control the workings of schools, then, it is feared, the schools will be controlled by industry which has a profit maximization motive rather than a societal good maximization motive [7, 16]. For example, many school systems have a policy that all children have a right to go to school. The policy is enforced by laws, attendance records, school procedures, teacher norms and practices such that all children, regardless of their native language, ability level or physical disabilities, are accommodated. These rules and regulations are viewed as positive measures for the society. If the rules did not exist, unregulated markets might well only serve the most intelligent, ready-to-learn children. Thus, educational systems have traditionally been funded, designed and controlled from using a top-down, bureaucratic system of bureaucracy within a government.

2.2 Expert Driven

Knowledge is viewed as an established, known entity. We can know, test and master chemical compounds, mathematical principles, poems, writing skills, history lessons, past tense of verbs, theories and the history of impressionist art. It can be found, mastered and tested. It is found in expert settings such as in texts, validated books (or web sites) and within teachers (or professors) [18]. Mastery of such knowledge is highly valued and is the primary goal of educational systems. Knowledge which does not fit within the validated, expert knowledge set is opinion, conjecture, unproven and untested. Wikipedia can never be a good source, nor can user groups because they have not been validated by an expert – anyone can post.

In this view, knowledge is static although it admittedly grows. But it grows by building, step-by-step, upon expertise [1]. Knowledge is not a flow, a system or a living thing. The knowledge that various streets are impassible through flash flooding, passed through twitter networks cannot be knowledge – no matter how true. It cannot be knowledge because it is not built from expertise. It is not validated.

Therefore it is conjecture or opinion. It may be high reliable conjecture or opinion, but, nevertheless, it is not knowledge [6]. Thus, collective intelligence cannot create knowledge unless the people behind that system are fully vetted as experts.

2.3 Outcome Defined

This type of knowledge should be mastered and success in mastery can be measured by individualized tests. Although some educational outcomes are broader than simply mastery of material, they are all generally defined by an individual mastering a set of knowledge, skills, analytic capabilities or ability to produce a standard measure of output such as a paper or thesis. This differs substantially from the goals of collective intelligence which is systems driven. The systems builds knowledge in real time, so knowledge changes rapidly. But, equally important is the system for collecting the knowledge. It must be quite open. When it is restricted, say by a political barrier or an institutional barrier, then the quality of knowledge deteriorates or does not stay up to date. The system must also be self-healing. When bad information is present, there must be a way to tag it, eliminate it and assure that it is less likely to appear in the future. Thus, the process of building knowledge is as important as the knowledge itself.

Contrast this to school-based knowledge which is assumed to be fixed at any given time. It changes not dynamically, but rather grows in a linear fashion. Pluto is no longer a planet – according to the experts. Twitter may be a social force, but the experts cannot agree on that. Only facts can be verified by experts so facts are knowledge. Facts and skills are testable. It might be useful for the society to have user groups, research networks, open source software and google, but the collective intelligence behind them does not have a place in schools. None of them establish validated knowledge and skills that can be measured. They are tools but not knowledge-generators or skill transmitters.

2.4 Efficiency Constrained

If outcomes can be strictly measured, then efficiency is determined by achieving those outcomes for the least expenditure of resources[3]. An efficient school might be one that educates children adequately for a low cost. It might be one that achieves high test scores within a reasonable budget. An efficient teacher might be one that successfully covers the curriculum in the time allotted for it to be taught or a teacher who keep students focused on their class work even while another teacher spends much time on classroom management. An efficient educational system finds ways of delivering educational content for lower costs without deteriorating test scores or serving fewer children. Outcomes are measurable – children served, time spent on school work, test scores, or curriculum covered. Costs are measurable – budgets, teacher time, resources.

In a system based on collective intelligence, efficiency is likely to measured as a flow – how useful is the system for learning? If one can learn a lot without spending too much time using it or adding to it, then it is efficient. Because the amount of learning depends upon the willingness of users to contribute, people work toward a common goal of learning even, if in the shorter run, it somewhat constrains their time resource. Building an efficient learning system is worth the time and energy if, over time, it becomes increasingly efficient. It gets increasingly efficient with more focused learners, collective intelligence systems of self-regulation and easier methods for sorting and using.

Since formal school systems are focused on knowledge that is expert-validated, learning per se, if narrowly defined as the ability to memorize or internalize known knowledge. Thus, time spent building a flexible, dynamic learning system designed for innovation or creativity cannot, almost by definition, be efficient. It may be fun, but not efficient. It does not lead to higher test scores, more time on curriculum or more children served.

3   Underlying Structural Differences

If the only difference between formal schooling views of knowledge and that implied by collective intelligence systems were their characteristics, the integration of collective knowledge into formal education would not require a complete redefinition of education. It would require retraining of management thinking starting at the top all the way down to the teachers, possibly the parents and, perhaps, the students. But the situation is not that easy.

The formal schooling characteristics derive from a particular structural logic. The logic derives, in large part, from an industrial model of expertise and control. Since mass education, and, in turn, the formal educational system emerged from the industrialization of the West, this is not a surprise [11]. Collective knowledge systems, on the other hand, manage flows (systemically) in order to allow outcomes the most fluidity (innovation) [8, 12]. Such systems have come into widespread use as technology has become cheap, in common use and linked globally and as knowledge innovation has become a major source of new value creation.

These structural differences mean that there is a stronger set of logic that grounds the characteristics of each system. That is, it is not just a matter of education, use of technology, or modernizing methods. If formal educational systems are to embrace collective intelligence, they will require a systemic revamp. To understand this, the two systems need to be detailed and contrasted.

3.1   Collective Intelligence Systems

Collective intelligence systems are a subset of intelligence systems and are best defined by their flow and linkage of components or nodes. The systems that drive

Wikipedia, Google or help direct software development are collections of people who reference each other's knowledge and contribute knowledge to the common pool. The collective grows and builds knowledge which becomes ever more sophisticated, easier to use and innovative. Thus, collective systems are best described by the process by which they create knowledge – learning.

The systemic flow of knowledge builds innovation because no one person need hold all the information. Unlike a scientist working in isolation in a lab, collective intelligence has the ability to take the particular knowledge of individuals, combine it and build a complex system. By combining individual pieces and working collectively to build a complex system, no one person is the final expert [2]. Each person holds a piece of necessary information and has the skills to work collectively.

It is also dynamic. If we search on Google for tsunami in December 2010, we will get some general information. If we do so in March 2011, we will get hundreds of web sites on Japan and many linked to nuclear concerns. The information changes as the collective learns, grows, changes direction, builds new understandings and demands new information. So information is largely treated as dynamic. There may well be unchangeable facts. But the use and context of information is constantly active and the most recent has the most weight.

Collective intelligence can also handle rapid change of knowledge. Because no one person or source is the ultimate authority, knowledge can and does change as each person learns from the system at hand. Knowledge is defined not by its definition at any given moment, but by its systemic qualities. No one expects to get the same search results from Google from one day to the next, but one does expect that the algorithm behind the search constantly returns the most relevant information. No one hopes that user groups define, at a single moment, the knowledge base for a given topic. But everyone expects a healthy collective knowledge system to be able to produce the best answer with relative ease. Knowledge is not finite, but the system of producing it has quality standards nonetheless.

Finally, many systems which build collective intelligence are often adaptive. When the system is not working exactly right, it self-corrects and stays alive. Wikipedia is largely reliable because people monitor it and report when the information is inaccurate. Collective intelligence usually swamps a few bad informants or has a way of isolating them against future participation.

3.2 Formal Education Systems

In contrast to intelligent systems, educational systems are generally defined not by the knowledge they create but in physical terms or policy structures. Schooling is a set of buildings, classrooms, teachers, texts, curriculum, desks and learning materials. It has a given structure, progression, standards of measurement for student progress, teacher quality, school success and regional progress. To regulate such a system, one begins by asking how many students there are which, in turn regulates how many teachers and classrooms there must be. Policies determine flow, content, definitions of quality, conduct, inputs and outputs.

These structures derive from an industrial era and a former technology. Industrial era schools needed to produce workers that had a standard set of skills – be they factory workers, managers, engineers or designers. The source of knowledge was largely pre-determined. In school, it was the teacher (or professor) and texts (or library). Once in the work world, one referenced back to these expert sources and the knowledge one accumulated and built on the job. Schools needed to be built such that each child could (1) access the school building daily from home, (2) be within earshot of the teacher, (3) have access to learning materials.

The logic of the industrial era also set the standard for the regulation of these schools. Much like factories, these giant systems of education are headed by a CEO (superintendent), governed by a board (school board), measured for success (tests, teacher qualifications, school building standards), and are regulated for quality control (curriculum, school managers, teacher in-service training).

These physical components and Taylorism-style management structures still define schooling systems even as technology begins to take hold in schools. Even though many schools have ICT systems, the usage of such systems is still organized around physical spaces, hierarchical governance, standardized tests, and learning measured by individualized mastery of knowledge. Knowledge is still treated as static and, learning is in service to knowledge acquisition – teachers teach, students learn.

In contrast, one would need to start far away from the physical structure of buildings and classrooms and the management structures of superintendents and school boards if collective intelligence were to be best used in formal schooling. One would have to start just where collective intelligence systems always start – what needs to be learned or built? Since collective intelligence works best when a problem is tackled, give the school a problem. Maybe it can take on a three year effort to improve the transportation system in the neighborhood to be greener. Now, build a learning system around that problem. Now, figure out which subjects can be built into the learning system. Math? Physics? Civics? Science? Writing? How can collaborative learning help students to solve the problem best? Then, the question is infrastructure. It now involves computers, cell phones, internet links, social networks as well as books, teachers and classrooms. Infrastructure is an outgrowth of collective learning environments rather than the starting point of formal schooling.

4 Underlying Economics

These structural changes are substantial, requiring not just retraining, but a complete overhaul of how the entire learning system works within formal schools. Nevertheless, if the stakeholders are willing to consider the possibility that schools are not necessarily buildings and student's primary goal is learning (rather than test performance), it is within the realm of possibilities. But it is well worth looking one layer deeper to ask the question of why these two learning systems have evolved so differently. If both formal schooling and collective intelligence are functional systems, why are their underlying systems so different? The answer reveals an even deeper shift in the underlying economics. Linking the economic changes to the shift in learning system logic provides some insights into the possibilities and also provides a reason for rethinking the system of formal schooling.

4.1 The Industrial Economy

Collective intelligence systems hit their stride when knowledge became a primary source of new economic value and when concomitant changes in technology allowed people living far apart to easily collaborate in near real time. But, formal educational systems grew out of an industrial era when nations where defined by their industrial strength and people needed training to move off the farms and into industrial towns [7]. In this era, education moved from a privilege of the upper classes to a national priority. If industry was to grow, people needed a basic set of skills which prepared them to work in industry.

Under such an economic system, innovation was not as important as mass production. A few innovators were vastly rewarded, but the real economic gains were to be made in taking an idea and moving it into the production stage and doing that production with increased efficiency. Machinery in industry, homes and civic life changed the face of cities and the ways that people live. The quality of life changed substantially and moved people from small farms to factories and, later, into centrally heated and cooled city apartments with televisions and automobiles. Mass production was the prime source of the jobs and lifestyles that creates this new value.

In such an economy, education prepared workers for the production line and for the specialized jobs of management, design and engineering. Teachers became professionals with like training and protocols. Schooling was a system from training to work in these regimented environments and provided you with a known, proscribed set of skills designed to equip you for a given job. Tests at the end of school assured that you did have the skills. Experts validated the knowledge and school administrators made sure the environment was conducive to the larger purposes of society which meant that all children had a role in this great industrial process. The systems of formal schooling are a direct outgrowth of this economic imperative as are its characteristics.

The problem for formal schooling is that the economic system has changed, in large part. Although factories still abound, especially in poorer countries, the new value that is being created has a different economic logic. Since it is the economic logic that underlies the system and the system that underlies the characteristics of formal schooling, the changes in economics likely spells an inevitable change in formal schooling. The emerging economic system support not so much static, expert-validated knowledge as it does collective, system-built innovation.

4.2 The Learning Economy

As technology spreads, the value embedded in the objects we seek, our cell phone, jet engines, and ATM machines, are not so much pulled from the earth as they are pulled from the brains of innovative, talented and creative people the world over. In advanced industries, mastery of knowledge, concepts and application skills continue to build year by year. Although new ideas, approaches and applications are the tangible outcomes of the knowledge economy [5, 21], the more intangible ways in which these new ideas form represents the systemic change in the economy [15]. Collective intelligence is a good example to illustrate the contrast.

A new smart phone may be the latest entry into innovation that generates capital using knowledge. The knowledge integrated into the phone is nearly all of its value - very little is in the plastic, copper and glass which is the physical form of the phone. If the phone sells for $400, then, perhaps only two percent of its value is in the physical product. The rest is knowledge. But the company which produces this phone has a very small window of time in which to realize these profits. That phone model will be outdated within months. Sales will turn to another design.

But the continuing value of the new design is in its integration of learning [10, 13, 14]. First, the team that designed it also increased its ability to learn during the design phase. Learning to work with each other, to understand who has which skills, which sub-group needs to meet when, how to use technology, time zones and communication software to enhance the design and its effective application. They may well have overcome language or cultural barriers to find commonalities in approach, interest, strengths and learning styles. By having a functional learning culture, they have a distinct advantage over a rival team at another company. The rival team will have the raw knowledge embedded in the product, but not the learning team culture. A learning culture takes times and gets more valuable as time goes by.

The new smart phone has another piece of learning value. Its integration of embedded learning systems means that the consumer is also buying future learning capabilities for the phone. The phone may be a dynamic learning instrument that can absorb new technologies (apps) and adapt to a changing environment. Thus, the phone's ability to dynamically integrate collective intelligence is also a selling point. Collective learning adds value to the product both by creating and building a more fruitful learning environment for innovation in design but also by embedding the phone with the potential to grow and adapt.

A school-wide project to find greener transportation alternatives for its community could pattern this value creation. It could create value (a new design), build collective intelligence (the ability for groups to learn) and embody collective intelligence in its product (design with collective intelligence integrated). But, formal educational systems cannot do this because they are expert-driven, based on a model of fixed intelligence and in which knowledge is the ultimately goal.

4.3   Value Creation

The chance to improve our lives through e-government, social networks, medical data files, email, global news networks reminds us that the word "industry" is out-moded to capture where value is created. Value is being created in all walks of life. Value is increasingly created throughout sectors of society – not just industry. Knowledge has a particular set of economic characteristics which stand in contrast to industrial economics [9, 17]. As such, collective intelligence is more than a useful tool within a knowledge economy – it is a product of a fundamental shift in the way that value is created. This is a shift not yet embraced by the bureaucratic systems that govern education. Understanding the new economic logic explains why formal education systems cannot easily adapt.

If learning has inherent value in a knowledge economy, collective intelligence has yet another use for value creation. It derives from another fundamental economic shift. The value of knowledge and learning has the potential to make all sectors more effective and cheaper. Productivity, per se, is no longer restricted to one sector. If e-government is more efficient than physical offices, innovation occurs to make it so. If awareness of alternative information systems makes populations uneasy to join a global system of flows, then civil society will adopt social networks to bring about political change. If adults need to learn new technologies, then learning systems will grow out of the internet and its strengths of collective intelligence.

The old economic theory that defined societal welfare largely as a result of pro-duction is eroding in explanatory value and robustness. OECD has undertaken a multi-year exercise to rethink how it measures societal progress since GDP is no longer an accurate measure of how much a society has progressed [9, 19]. Some knowledge production does not even take the form of monetary value such as the value created by Google or Facebook. Much of their value is never counted, taxed or aggregated. But the major value gains occur because multi-sector, globally inte-grated, complex, adaptive, dynamic systems must be managed, counted, regulated and searched for weaknesses. The economy has spawned an industry based on sys-tems management, innovation, design, monitoring and regulation that does not fit the older definitions of production. And yet, it creates substantial value. Systems of climate, disease, food, finance, security, knowledge, transportation, education and communication bring large value to our lives but also represent threats as a weak-ness in one part of the world affects us all. This systems industry is complex, in-volving private business, government, volunteers, civil society, churches, physical and virtual networks, knowledge which goes to market and knowledge which never has monetary value [23, 24].

5   Conclusion

Today's value cannot be built, innovated, monitored or fixed using expert models of intelligence or static views of knowledge. It requires collective intelligence; yet, the currently formal education system contains few tools to prepare students to work with these systems other than the field of System Engineering itself. The collective knowledge skills which are endemic in this emerging skill-set are not taught, practiced, assessed, valued or grown within the formal education system.

This paper began with a question of whether the lack of integration of collective intelligence systems in education was due to training or attitude deficiencies as has been suggested in the literature. It traced this resistance to a larger philosophy of formal schooling and its way of managing knowledge. But this philosophy was linked to a system of controls, management and policies which are interlined and cohesive. This system could then be traced to an economic era from which it derived. Collective intelligence, although a technical skill and attitude, also has an underlying philosophy, systemic logic and is founded in a particular economic era. The conclusion must be reached that formal schooling failure to integrate collective intelligence derives not so much from attitude or technology skills, but from system controls which served (and continue to serve) a particular economic system – even as it fades in influence.

But formal schooling is unlikely to go away because young children do need basic skills. Possibly those skills ought to be rethought in terms of learning skills rather than static skills, but the foundations must be laid at an early age. This early age is also a time when schooling serves as a babysitter and children cannot generally be productive otherwise. Thus, schooling is likely to continue in some form. But it is equally likely that the economic logic of today will impose itself into private or quasi-private systems that allow more creativity, flexibility and experimentation. Government control of such systems will (by necessity) be relaxed and collective intelligence will be integrated. It is not the inherent appeal of collective intelligence, but its clear superiority in producing people who are good learners, innovators and collective thinkers that will cause the change. The logic of schools was established by an industrial system of production. The new learning economy will provide the framework for integration of collective intelligence.

References

1. Apple, M.: Official Knowledge: Democratic Education in a Conservative Age. Routledge, London (2000)
2. Axelrod, R., Cohen, M.D.: Harnessing complexity: Organizational implications of a scientific frontier. Basic Books, New York (2000)
3. Cohn, E., Geske, T.: The Economics of Education. Butterworth-Heinemann, Maryland Heights (1990)
4. Cooper, B., Ehrensal, P.A., Bromme, M.: School-Level Politics and Professional Development: Traps in Evaluating the Quality of Practicing Teachers. Educational Policy 19(1), 112–125 (2005)
5. Cortright, J.: New Growth Theory, Technology and Learning: A Practitioner's Guide. Impressa Inc., Portland (2001)
6. Demetriadis, S., Barbas, A., Molohides, A., Palaigeorgiou, G., Psillos, D., Vlahavas, I., Tsoukalas, I., Pombortsis, A.: Cultures in negotiation: teachers' acceptance/resistance attitudes considering the infusion of technology into schools. Computers & Education 41, 19–37 (2003)

7. Deyoung, A.: Economics and American Education: An Historical and Critical Overview of the Impact of Economic Theories on Schooling in the United States. Longman, London (1989)
8. Engestrom, Y.: Training for change: new approach to instruction and learning in working life. Working Paper. International Labour Organisation, Geneva (1994)
9. Giovannini, E., Hall, J., Morrone, A., Giulia, R.: A framework to measure the progress of societies. OECD Working Paper, Paris, http://www.oecd.org/ (retrieved from 2009)
10. Hage, J., Meeus, M.: Innovation, Science, and Institutional Change: A Research Handbook. Oxford University Press, Oxford (2009)
11. Hartley, D.: Tests, tasks and Taylorism: a Model T approach to the management of education. Journal of Education Policy 5, 67–76 (1990)
12. Hyysalo, S.: Learning for learning economy and social learning. Research Policy 38, 726–735 (2009)
13. Kim, L.: Learning and Innovation in Economic Development. Edward Elgar Publishing, London (1999)
14. Lundvall, B.A., Johnson, B.: The Learning Economy. Journal of Industry Studies 2, 23–42 (1994)
15. Lundvall, B.A., Rasmussen, P., Lorenz, E.: Education in the Learning Economy: a European perspective. Policy Futures in Education 6, 681–700 (2008)
16. Marginson, S.: Markets in Education. Allen & Unwin, St. Leonards (1997)
17. Mirowski, P.: Why There is (as Yet) no Such Thing as an Economics of Knowledge. In: Kincaid, H., Ross, D. (eds.) The Oxford Handbook of Philosophy of Economics, pp. 99–156. Oxford University Press, Oxford (2009)
18. Moore, R.: Professionalism, Expertise and Control in Teacher Training. In: Wilkin, M., Sankey, D. (eds.) Collaboration and Transition in Initial Teacher Training. ch. 2. Routledge, London (1994)
19. Organization for Economic Co-operation and Development. Measuring Progress. Third OECD World Forum, Korea (October 2009), http://www.oecdworldforum2009.org (retrieved on June 10, 2010)
20. Reeves, D.: The Leader's Guide to Standards: A Blueprint for Educational Equity and Excellence. Jossey-Bass, New York (2002)
21. Romer, P.M.: Two strategies for economic development: using ideas and producing ideas. In: Klein, D. (ed.) The Strategic Management of Intellectual Capital. ch. 13, pp. 211–238. Butterworth-Heinemann, Boston (1998)
22. Stiglitz, J., Sen, A., Fitoussi, J.P.: Report by the Commission on the Measurement of Economic Performance and Social Progress, http://www.stiglitz-sen-fitoussi.fr (retrieved June 10, 2010)
23. Taylor, M.C.: The moment of complexity: emerging network culture. University of Chicago Press, Chicago (2001)
24. Thelen, E.: Dynamic systems theory and the complexity of change. Psychoanalytic Dialogues 15, 255–264 (2005)

On Presence, Collective Performance and Assumptions of Causality

Mark McGovern

Abstract. The ways we assume, observe and model *"presence"* and its effects are the focus in this paper. Entities with selectively shared presences are the basis of any collective, and of attributions (such as "humorous", "efficient" or "intelligent"). The subtleties of any joint presence can markedly influence potentials, perceptions and performance of the collective as demonstrated when a humorous tale is counterpoised with disciplined thought. Disciplines build on presences assumed known or knowable while fluid and interpretable presences pervade humor. Explorations in this paper allow considerations of collectives, causality and the philosophy of computing. Economics has long considered issues of collective action in ways circumscribed by assumptions about the presence of economic entities. Such entities are deemed rational but they are clearly not intelligent. To reach its potential, collective intelligence research needs more adequate considerations of alternate presences and their impacts.

Keywords: Presence, humour, causality, collective performance, economics.

1 Introduction

An assumption apparent in much discussion of collective intelligence is that "together is better", but this is not necessarily so. After all, history is replete with collective failures. Collective issues lie at the heart of much social theorising, probably most famously in the "problem identification" of Karl Marx and the various "solutions" since offered which have shaped humanities lot. Noted here are the more conceptual aspects of the works of Olson, Chandler and Coase in economics who considered problems of "collective" attribution, action and achievement. Though

Mark McGovern
Queensland University of Technology
e-mail: m.mcgovern@qut.edu.au

J. Altmann et al. (Eds.): Advances in Collective Intelligence 2011, AISC 113, pp. 103–114.
springerlink.com   © Springer-Verlag Berlin Heidelberg 2012

paths differed, all essentially responded to the broad question of *"What causes collective successes?"*

To begin the development, a tale is considered with an implicit context of "What causes it to be humorous?" While the structure of the tale and the attributions made to story elements are part of the story, *fluidity in interpretations* appears to be a key element. This appears especially important when the listener revises interpretations, contrasts alternate views and/or enjoys the resolution of dissonances.

A simple schema is used to outline and explore a range of "possible presences" and how these might be alternately demarcated, joined and interpreted as "a collective". In a humorous tale, a collective of interpretable entities are *assumed alternately present* during the relating of a tale. Alternate presences of the same entities take on importance as path, node and context dependencies unfold with distinctive impacts.

Such a multiplicity of presences associated with an entity stands in sharp contradistinction to the convergent, determinate approaches widely favoured today, be these styled in mechanistic, statistical, formulaic, or habituated terms. Functionally, a better logic is needed, and an *aor* (and/or) relational offers a promising if frustrating development (as is discussed in a companion paper to this one). Dynamically, an impact approach opens doors that other approaches close. More ambitiously, we need better paths through the complexities of indeterminate experiences, a more apt metaphor for the actualities of life, a more enabling thinking framework, and a deeper philosophy of computing. These later aspects are a focus in this paper though the argument remains discursive and many relevant aspects are only touched on lightly.

In all, a deeper appreciation of intelligence is sought. An improved ability to express alternate configurations and their impacts is a goal. Exposition touches on several aspects that together offer a trail towards improved understanding of collective intelligence. Humour provides a rich setting for our explorations. There are clear challenges. Those working as investigators and arbiters of collective intelligence can do much good if the issues and challenges presented are well addressed.

2   Learning Anew from an Old Joke

Humour[1] is one of the hallmarks of humans and, arguably, intelligence. It may be found in many tales, such as:

> What is the definition of Heaven? A French cook, Italian lover, English policeman and German engineer.
> And of Hell? An English cook, German lover, Italian policeman and French engineer.
> And of Subprime EUrope? An Irish banker, Greek statistician, Spanish investor and Portuguese administrator.

---

[1] The alternate spelling of "Humour" with a "u" is used here to assist in highlighting an aspect.

Consider your response to this ordered selection of words: affront, mirth, indifference, curiosity, boredom and/or something else. What is the intensity of your response? What are the elements of your response and which attributes or dimensions would be important in you articulating "my response" – and how might the perceived motivation of the teller influence your position and its communication to the teller or an other? In which ways is this response shared, by some others or collectively? How, also, would your response vary if the nations were from Asia, Africa or the Americas?

Responses to a common stimulus can vary widely, for all manner of reasons. The nature and intensity of responses is based upon some selected attributes, predilections and perceptions with communications particularly conditioned "both ways". Importantly with humour, the real *intent and meaning lie beyond the text* with any literal translation missing the whole point of the tale. The "u" in humour is pivotal, perhaps *via* "you" or "U2+" (encompassing "us both", "we together" and the like).

Areas with potential in researching humour and "collective intelligence" include:

- roles of humour. *Why do humans tell a joke, and do other entities share a sense of humour?* The pervasiveness of humour points to a range of roles as well as strong motivations.
- humour and intelligence research[2]. *Does humour reflect or require intelligence of some ilk?* Links between humour and intelligence (both natural and artificial) though long discussed remain speculative [2, 7, 12, 14]. While intelligence of the producer, teller, listener and peer group can be variously considered, much humour does rely upon sufficient prior shared knowledge, a recognition of parody, "knowledge about knowledge" [8] and knowledge of the world [12]. Some minimal intelligence and knowledge do seem to be needed if humour is to be appreciated, especially when complex meanings are implicit in the text or tale. Interpreting text itself is an important and ongoing challenge [10, 11, amongst others] which lies beyond the current considerations.
- causes of mirth. *Why is something seen as humorous by an audience, be it of one or many?* Surprise is one important element of humour: what was expected does not result and the pleasant irony of our situation (as we "clear thinkers" are tricked) evokes enjoyment and mirth.
- repetition. *Why is repetition common, how does it affect enjoyment, and how does it use memory and context sensitivities?*

  - Repetition *in* the tale is here a key structural element that allows simplicity and reinforcement of a particular line of development.
  - Repetition of *"signature parts"* of a tale predisposes the listener. *"What is the definition. . . ?"* may not be as familiar as *"Why did the chicken. . . ?"* or *"Knock, Knock."* but each quickly orients the reader as to the nature and anticipated impacts of the tale.
  - Repetition *of* a tale may reduce enjoyment, but repetition may be part of long-term bonding where the lack of novelty is itself a sign of familiarity and a source of amusement to those involved.

---

[2] The considerable literature in this area can only be touched on very lightly in this paper.

- aspects in the tale itself. *What groupings, relations and stereotypes are used and in what ways are they important?* In our tale, nominated parties are assigned particular roles; role reversals and associated dissonances mark the difference between Heaven and Hell; and a new twist (using the now twice-related structure) reflects onto a contemporary situation.

The roles, nature, impacts and formulations of humour are indeed many and varied.

However, is it something not yet mentioned that "really makes the humour work": *the supposed presence of some quality in each nominated entity and associated evaluations by listeners?* The tale engenders mirth due to associated but hidden presences and implicit processing by the parties communicating. Combinations offered are sufficiently plausible and mutually supportive to allow easy acceptance, especially since humour involves playful fun, whatever the intents of the parties.

Suggestiveness and interpretability subsist in the tale[3].

- *via structure, for example.* A humorous tale is typically concise or punchy with a reliance on structures. The order of parts matters as each adds something new while selectively building on what went before. A break in a telling of the tale sees a loss of momentum, framing and impact.
- *via fluid innuendo.* A meaning may be suggested but not defined. A meaning need not persist across the life of attribute, entity, tale or event.
- *via anticipation.* Expectation builds of something more. In the tale, some may have appreciated "the hint" of an extension, to a further part for *EU*rope with not just some better mix of elements but also new (Union) elements. Whether this would be towards Heaven or Hell remains open.[4]

The text itself is deliberately underdefined so as to allow flexible positioning and more active or imaginative engagements.

Positioning occurs during both the setting of the episode and the relating itself. Telling the tale at a funeral service would be in bad taste but may well be appropriate at a wake afterwards. A teller who interrupts to begin, stumbles on parts and rushes to the end can make even the best tale forgettable. Warm engagement outshines serious demeanour, and so on. Such things are contextual. They transcend the tale itself *externally* but can be pivotal in the development of good humour. Such things involve external referents and presences beyond the tale.

Internal transcendence involves presences implicit in the tale itself that provide both connectives and bases for evaluations that make the tale work. Qualities such as fairness, creativity, passion and spontaneity as each assumed variously associated

---

[3] *Reducing* such an ability "to manage to survive" is, of course, a goal of clear definitions and objective specifications. Hints, suggestiveness and alternate interpretations are *not* welcome in formal analysis, contracts, well-disciplined discussions and the like. Considerable efforts are made to reduce ambiguity, irrespective of whether any ambiguities persist in actuality.

[4] The suggestive big-collective "U" (as in EU or European Union) is able to enter explicitly here due to a property of print and also alphabet correspondence of symbols "u" and "U". Such an artifice would not be available in an oral presentation, reflecting how the medium influences the telling of the tale.

(as in the "fair" "English police" of "Heaven"). The assumed transcendence of qualities[5] which are also constitutive of "the broad state" (of Heaven or Hell) holds the tale and evaluations of it together. Those making literal interpretations or those not adept at using fluid idealisations would find the whole thing confusing.

3   Bundling Roles and Questions in a Collective

The structure used in our tale is a simple one: pick four activities; assign four actors firstly according to a stereotype and secondly via some outrage on the stereotype; thirdly use some switch while maintaining four roles and four actors, while hinting at some further presence as a basis for a fourth stage.

Patterning occurs when assembling any collective, be it for humorous or "more serious" purposes. The frame in use is illustrated in Table 1. Four main areas are:

- " an encompassing outcome, here *selectively* instanced as "Heaven", "Hell" and "Subprime Europe", or H_1, H_2 and H_3.
- an active actor, a, specifiable by *successive selective* drawings (without replacement) from a *chosen* set of actor A.
- a qualifier nationality, n, specifiable by *successive selective* drawings (without replacement) from a *chosen* set of nationalities N.
- a select combination area comprised of four combinations, each having been drawn *deliberatively* from A and N.

The "whole joke" seems captured in such a table, but have we captured the "essence of (the) humour" or why some combinations are "funny" and are so regarded, or not?

A table such as this is *a representation* that could be used further, or it could be simply appreciated as capturing a description efficiently. For example, the table could be passed around at a party but confusion rather than mirth would be likely (unless, perhaps, the party goers are "matrix obsessives" at a convention). The new medium, in interposing itself and its peculiar properties between the parties, loses (at least much of) the message while the proffering party misses the intent of opening rich lines of communications between persons.

"Welcome", we may say, "to email and the world of modern communications, management and the like" where the medium constrains the message and messages in a minimal medium abound. Text messaging appears cheap to users but the frequent need for multiple messages indicates a limited quality of communication. Such messaging is excellent for some purposes but inadequate for others. "Fit for purpose" involves a whole range of evaluative factors.

Suppose we were to now subject Table 1 *in isolation* to exhaustive analysis. Placing it in a "suitably secure" environment we would "expertly apply sophisticated methods" to determine the properties of the table *and* (we assume) the key to

---

[5] Which exist and can persist beyond and/or within their hosts.

Table 1 Framing a tale

Situation		Outcomes H	Agents			
	2 in	[Hell]	1	2	3	4
			cook	engineer	lover	policeman
Nationalities	1	French	[Heaven]	[Hell]		
	2	Italian			[Heaven]	[Hell]
	3	English	[Hell]			
	4	German		[Heaven]	[Hell]	
			Key: Outcome H	Inputs I		
			Heaven	[Heaven]		
			Hell	[Hell]		

humour.[6] Alternately, we may seek a very comprehensive knowledge base that is *assumed* all-encompassing with the keys to humour "in there some where".

Have we now "lost the plot"? No matter how energetic our efforts, skilled our application, noble our intents or well-informed our actions, we have embarked on a risky path in our quest for the key to humour. What considerations apply to collective intelligence research? Tellingly, Coase notes the divorce of economic theory from its subject matter as pivotal in the transferability of "the economic method" to other disciplines and their "rejuvenation". The result is expressed in his famous statement: *"the rational utility maximizer of economic theory bears no resemblance to the man on the Clapham bus or, indeed, to any man (or woman) on any bus."* [9].

4 Dissecting a Frog...

"Explaining a joke is like dissecting a frog. You understand it better but the frog dies in the process"[7] [15]

---

[6] For example, We could posit optimal cell patterns, perhaps diagonalising the cells of Heaven and analyzing the suboptimality (and Hellishness) of other patterns, perhaps in terms of distance from the diagonal (as clearly Heaven is "best practice"). The results of such analysis might then inform "scientific management" schools, "human resource" practices and all manner of theories.

[7] E.B. White as at
http://tvtropes.org/pmwiki/pmwiki.php/Main/ptitle0t9r68ih.
Several variations can be found including:

- "Analyzing humor is like dissecting a frog. Few people are interested and the frog dies of it." http://www.quotedb.com/quotes/704
- "Humor can be dissected, as a frog can, but the thing dies in the process and the innards are discouraging to any but the pure scientific mind." *Some Remarks on Humor*, preface to *A Subtreasury of American Humor* (1941) http://en.wikiquote.org/wiki/E._B._White

Each has differences, including as to its essential meaning and moral. The last appears original.

Explaining humour is something of a thankless task. Yet it is an important challenge, particularly if humour is regarded as *"by far the most significant behaviour of the human mind."* [3]

Interestingly, each listener can bring can bring distinctive responses as they enjoy the tale. Consider, for example, the need to *consonate*[8] and *concatenate*[9] Heaven, English and police in our tale. The three words need to not only be linked in some way but also to "sound in sympathy". Linking by juxtaposition within a simple repetitive structure is easy but "consonation" plays on unstated, underlying presences presumed known by all parties.

If pressed to explain such a consonating presence, a listener might return a word like "fair", but in listening to the tale he or she may not have explicitly identified any word or phrase. "Reliable" or "neutral" or "polite" could work as well as "fair" for example for various listeners. The "essential presence" that allows the combination to appear plausible is not only undetermined but also variable across participants.

In moving to Hell, the original attributions[10] alter as both "English" and "police" are reassociated in a new and polar-opposite setting. Something adverse about "English" citizens is supposed revealed when they are in a kitchen, and similarly "an Italian" is supposed ill-suited to a policing role. Whatever "essential presence" each carries, "it" works to good effect in some situations and to bad effect in others. Alternately if multiple attributes are allowed, those supposedly evidenced in one setting are more beneficent than those evidenced in another. So we allow a presence with differentiated effects depending on context or, alternately, a mix of presences which are alternately displayed in one context or another.

There is an important further consideration: that a group attribution need not apply to all individuals, or to some individuals for some of the time. While there are many fair English police, there are some who are not and some who may only act unfairly at some time(s) during their career. Also, I am sure there are superb German lovers and fair Italian police, but I know none and have observed few. I lack evidence to challenge the stereotype (or class attributes) but in humour I recognize that the attributions are not meant to be taken seriously. The speciousness of both the particular attributions and their generality across individuals and times is recognized.

We are now deeply enmeshed in problems bordering *quiddity* and *haecceity*[11], words (largely) vanquished from modern times but ones inseparable from medieval theorizing as to the essence of "this thing", of what "this" shares with "that" thing, of why a man can do both right and wrong and how "an English" (or other so-defined) man could be in both Heaven and/or Hell. *Unless* we allow sheer randomness and so

---

[8] Consonate: to sound in sympathy.

[9] Concatenate : to link or join together as in a series or chain.

[10] Whether explicit or implicit, idiosyncratic or shared.

[11] *quiddity* involves the real nature or essence of a thing or that which makes it what it is; *haecceity* involves what makes some thing "this thing" allowing a unique identity. These are answers to the questions "what makes it?" (or "of what is this thing essentially composed?") and "what makes it it?" respectively. Such presences need not be always present.

embed chaos in the heart of Man and caprice in the actions of men and women[12] OR determine that objects, including perhaps humans, can only have definitive properties that persist over time and place (mechanics following Newton), there will be a necessary indeterminateness in at least some properties. The grand OR arguably captures the foundational preferences of postmodernism and modernism along with their ability to determine "solutions". Otherwise, uncertainties evident *in situ* may be variously important and influential, or not here or now. *"Non-essents" can matter.* Intelligence appears to involve appropriate recognition of such things.

These three approaches to collectives outlined can be considered as:

- *"calm collectives"*. Groups or collections that are chaotic at the individual level but well-ordered in aggregate. There is some consistency or constancy in aggregate, overall or collectively which can condition "elements". The whole is sufficiently influential on the parts to ensure global orders are maintained.
- *"reliable objects"*. Objects that are well-ordered both in isolation and aggregation. Some individualisable attributes (are assumed to) persist and provide a basis for extensive orders. Some orders are both associated with an individual (set) and external-to-such-an-individual in effect.
- *"uncertain entities"*. Such entities are somehow-ordered and ordering in various situations (and perhaps repeatedly or consistently so) but they are not a priori nor situatedly determined or determinable. Rather order "occurs in situations" or impact events for reasons or via causes that may not be apparent.

These reflect the three great types of dynamical theories that underlie Western philosophy [13]: the all pervading influence or plenum; the unextended centre of influence as in the particle or force of Newton; and the interacting many of impact analysis. These can be symbolically expressed as $]1, 1]$ and $1]1$ where the "1" is the definitional unity in each case[13] and the "]" portrays an "encompassing"[14] hierarchical relation. This representation is further developed in another place.

5 "Collective Choice" in Economics, and Elsewhere

Considerations of *choices and their* various individual and collective *impacts* are central to much effort in economics. This is so not only at the individual level but

---

[12] Typically using a statistical artifice to allow well-behaved collectives with "likely" outcomes then conveniently deemed "almost certain".

[13] the unit "1" is effectively an expression of whatever quiddity might present in the situation under consideration. It requires further assumptions for the unit to be assumed as the foundational existence or a universally present manifestation. Haecceity could be acknowledged by qualifying the unit, as in 1_i or $1_\#$ where i is a set-order indicator and # involves some instancing (or partial revelation).

[14] the square bracket "]" is encompassing to the left (which is the side enclosed) and excompassing (or moving to things somehow beyond the individual realm) to the right. The bracket demarcates space into "here" (with the unit) and "there" (away from it).

also in aggregate, at the collective level. However, there has been a particular stylization that has become conventional which essentially assumes away problems of aggregation. The unresolved and arguably irresolvable gap between short run and long run considerations in microeconomics is testimony to this as are the distinctive differences and gaps between macro- and micro- economics. The development of a "bridging" meso-economics has been long neglected while studies in industry, institutional and regional economics have been regarded as peripheral or worse.

Ways by which grouped individual actions blend to collective outcomes are a particular focus of Olson while inadequacies in our notions of choice concern Coase. Chandler views the "visible hand" of the firm as instrumental in the operations of a market rather than the fabled "invisible hand" in its various guises. Consider these comments from each in turn:

- To deliver overall optimal aggregate outcomes Olson [9] considers explicit groups and their collective impacts. His foundational position is that: *"unless the number of individuals in a group is quite small, or unless there is some coercion or some other special device to make individuals act in their common interest, rational self-interested individuals will not act to achieve their common or group interests. ... The notion that groups of individuals will act to achieve their common or group interests, far from being a logical implication of the assumption that the individuals in a group will rationally further their individual interest, is in fact inconsistent with that assumption."* (p 2)
- For Coase [5], the posing of choice in economics results in *"the divorce of the theory from its subject matter ... the entities whose decisions economists are engaged in analyzing have not been made the subject of study and in consequence lack any substance. The consumer is not a human being but a consistent set of preferences ... We have consumers without humanity, firms without organization, and even exchange without markets."* (p 3)
- While many rely upon an invisible hand in a free market to monitor and coordinate economic activities, Chandler [4] instead sees the "visible hand of management" as actively structuring outcomes. Effects are uneven across sectors. *"These industries, where the visible hand of management had the greatest opportunity to increase productivity and reduce costs, were the most critical to the current health and continuing growth of the rapidly industrializing American economy."* (p 371)

Reflecting, hints to the development of a subprime EU appear. Unrecognised inconsistencies of "choice" (Olson) allowed untenable collective impacts to build while many visible hands (Chandler) were distractedly working on EU "Dreams" (*château de cartes*) and economic analysis was rendered impotent by its divorce (Coase), Hints as to resolution also appear, but their full consideration must be left to another place.

Several view of "causality" can be derived from such statements including:

1. informed

   a. *individual knowledge is a sufficient cause.* Individual rationality leads to, or can lead to, optimality overall;
   b. *promotion of a common view is a necessary influence.* Coercion or suitable devices are needed to direct individuals towards realising a common interest;

2. behaviour influencers

   a. *preferences alone cause behaviours.* Preference sets suffice to explain behaviours, there being no other matters of substance;
   b. *a mix of substantial influences cause behaviours.* The setting and substance of decision makers also matter in the making of decisions and associated behaviours;

3. performance

   a. *self direction delivers optimality.* Freely-associating, suitably-informed and self-interested individuals can self direct so as to achieve optimal outcomes;
   b. *direction by others delivers optimally.* Management by professionals drives performance.

These are not complete representations of any author's position nor the only formulations possible but propositions that might be tested in various circumstances.

A desire for a single line of argument or explanation runs strong in many people seeking to make order out of life experiences. Holding two lines of explanation simultaneously may broker plausibilities but it introduces uncertainties. Avoiding uncertainty, an analyst may prefer a single line in search of determinable solutions. Special solutions can result but their applicability may be (very) limited.

"The long concatenations of simple and easy reasoning which geometricians use in achieving their most difficult demonstrations gave me occasion to imagine that all matters which may enter the human mind were interrelated in the same fashion." Descartes as cited [6, p. ix]. Integration of the sciences by logical methods was "Descartes' Dream" (op cit), with Mathematics as the universal language and deductive logics serving as the means of connection *and determination.* Long chains of mathematical forms were to mirror actuality and reveal the Realities of our World and a wider Universe. All six "causes" above share this Dream.

However, and as was raised centuries ago in objection to the Dream, what happens when "an 'irresistible force' meets an 'immovable object'"? Each reflects a particular analytical position, a distinctive dynamic and a specific basis. One focuses on a tendency to change *position* while the latter tendency is to maintain "it". Each special view leads to chaos or stasis respectively. A binary divide is introduced when "all" or "none" *are assumed to* change. Any solution is then determined by a preference for 'the irresistible' or 'the immovable', and the relegation of the other view to an "other" universe of discourse or analysis.

In life, however, *some things remain while others change.* So some neglect life. By employing analyses where certain outcomes depend only on constancy assumptions and notions of cause used within formal models and some great dynamical theory, they build dreams that hold sway over the minds and lives of many. Such attitudes and positions present major challenges. As a more specific challenge, in seeking to program a computer, swarm, agent and the like to appreciate expressed humour *you may need to explain the joke to the machine,* to not only dissect the frog but to put it back together alive in another language and medium.

The wider issue is how we might move towards a more adequate philosophy of computing, one that can not only accommodate humour but also help guide "wise" (or at least "more sound") choices. The ways we use and develop "intelligence" in computing will be pivotal and potentially highly influential in societal progress.

6 Conclusion and Challenges

Problems arise in *any "being together",* as do potentials. Interdependencies exist and entities cannot be assumed to be fully analyzable as if they were in isolation. This is the practical experience of couples, families, firms and any organization of distinct parties or interests that adopt positions of interdependence. Each may have their motivation but the common assumption since Adam Smith has been that engagement is willingly undertaken where there is sufficient self-interest served.

Newton and others have provided a compelling model of analysis predicated upon certainty, clearly observable and stable values, explicit properties and the like. This powerful method resting with the third of the great dynamical theory types has allowed Descartes Dream of encompassing unification to be massively extended but perverted in the process. The global financial situation embodies such an extended perversion. The Dream is well suited to beams and stable bodies or the remoteness (from influence) of space but ill suited to the complexities of life, the dense interdependencies of uncertain living and the fluid interpretations used in humour.

What may be said of solutions for subprime Europe? We may offer alternate fourth parts to our tale. Some suggestions of principal actors (and actions) *to restore* European prosperity (perhaps a recoverable Dream of Heaven in EU on Earth?) are:

- *An Irish taxpayer, Greek worker, German investor and French depositor.*
- *A continent-wide bank, Commission statisticians, restored investors and administrators taxing anywhere funds can be raised to meet "obligations".*
- *Rebalanced accounts via repositioned bankers, investors, administrators, taxpayers and borrowers.*
- *A suitable share of $100 trillion over the next decade globally.*[15]

The real but tragic joke is that three such "Solutions" are being taken seriously. The arrival of previously uninvolved parties adds new complexities, but their potency

---

[15] a "Stop Press" late suggestion direct from Davos in January 2011. The WEF and McKinseys consider that since global credit on issue doubled between 2000 and 2010 from $50t to $100t it will need to double again in the next decade.

is assumed rather than reasonably established. While elements of each "Solution" may have merit, *ignored in all but the third is an underlying presence, that of unrepayable debts.*[16] Ignoring such underlying presences because they are not easily formalisable, explicit or definitively certain just perpetuates folly. What lessons can be drawn by those researching collective intelligence, or the lack of it?

References

1. Berger, A.: What's So Funny About That? Society 47(1), 6–10 (2010)
2. Binsted, K., et al.: Computational Humor. IEEE Intelligent Systems 21(2), 59–69 (2006)
3. de Bono, E.: I am right - you are wrong: from this to the new Renaissance: from rock logic to water logic, Penguin, Harmondsworth (1991)
4. Chandler, A.D.: The visible hand: the managerial revolution in American business. Harvard University Press, Cambridge (1977)
5. Coase, R.H.: The Firm, the Market and the Law. University of Chicago Press, Chicago (1988)
6. Davis, P.J., Hersh, R.: Descartes' Dream. The world according to mathematics. Pengin, Ringwood (1986)
7. Mihalcea, R., Strapparava, C.: Learning to laugh (automatically). Computational models for humor recognition. Computational Intelligence 22(2), 126–142 (2006)
8. Minsky, M.: Jokes and the Logic of the Cognitive Unconscious. In: AI Memo No. 6031980. MIT, Boston
9. Olson, M.: The logic of collective action: public goods and the theory of groups. Harvard University Press, Cambridge (1965)
10. Raskin, V.: Semantic Mechanisms of Humor. Reidel, Boston (1985)
11. Raskin, V., Nirenburg, S.: Ontological Semantics. MIT Press, Cambridge (2004)
12. Ritchie, G.: Current Directions in Computational Humour. Artificial Intelligence Review 16(2), 119–135 (2001)
13. Russell, B.: An outline of philosophy. Allen and Unwin, London (1927)
14. Stock, O., Strapparava, C.: Getting Serious about the Development of Computational Humor. In: IJCAI 2003 Proceedings of the 18th International Joint Conference on Artificial Intelligence, Morgan Kaufmann Mathematics, Ringwood (1986)
15. White, E.B.: Some Remarks on Humor. Preface to: A Subtreasury of American Humor (1941)

---

[16] Couched as loan contracts these have been on-sold on the basis of guaranteed income streams in "AAA" investments. Contracts and financial arrangements that strained against economic realities at inception are even less plausible now. Parties entering agreements do have responsibilities, but continuance is arguable when contracts are deeply flawed *and could have been, and sometimes were, recognized as such at the time of issuance.*

Preservation of Enterprise Engineering Processes by Social Collaboration Software

Dominic Heutelbeck

Abstract. In design and engineering, it is important to preserve more than just the actual documents making up the product data. For knowledge-heavy industries it is of critical importance to also preserve the *soft knowledge* of the overall process, the so-called product lifecycle. The idea here is not only to send the designs into the future, but also the knowledge about processes, decision making, and people. In order to preserve this knowledge, it needs to be captured at content creation time, a process currently mostly independent from the act of preservation. This paper discusses how to make tools and applications used at content creation time, especially in design and engineering, but also, in general, preservation-aware by using the OpenConjurer approach and framework.

1 Introduction

For innovation driven industries like design and engineering digital preservation is an important challenge. Legal regulations and contractual requirements often make the preservation of product data for decades mandatory. Product data is all data generated during the product lifecycle, e.g., including text, presentations, simulation data, and designs. The reuse of product data and knowledge in a rapidly evolving technological environment is both a goal and a difficult problem. The motivation for digital preservation in design and engineering is discussed in detail by Heutelbeck et al. [8, 9]. A key motivation is to preserve and reuse the knowledge about the actual design and decision making process. This includes documentation about who did what with the product data. The 'who' part of this information is called social context and not only consists of information about a unique identity of an actor, but also about his role and position within the social network of the people

Dominic Heutelbeck

Forschungsinstitut für Telekommunikation - FTK e.V., Martin-Schmeißer-Weg 4, 44227 Dortmund, Germany

e-mail: dheutelbeck@ftk.de

J. Altmann et al. (Eds.): Advances in Collective Intelligence 2011, AISC 113, pp. 115–130.

and organizations involved in the creation of the product data. The 'what' part of this documentation is called collaboration context and covers operations on the actual data as well as communication and collaboration surrounding the data, such as decision making, i.e., why something was done or decided. For instance, it could be important to understand why a component in a design was chosen over an alternative. This contextual metadata is required to answer potential questions of future users, who access the product data within the live environment, or later by accessing a preservation system. The following questions came up repeatedly during our interviews and discussions with potential industry users and experts:

1. Forensics and analysis of the past

   a. Reasons for commercial or technical failure?
   b. Which decision or negotiation process led to a fault?

2. Knowledge reuse

   a. How can we improve our products?
   b. How can we improve our processes?

3. Design reuse

   a. How can we enable a complete reuse of original designs in a new design?
   b. How can we reuse functional parameters for reengineering or maintenance?

Especially for questions 1a, 1b, and 2b the preserved knowledge about the processes and the social context of the content creation is important to both find the appropriate content, as well as for achieving the goals of the users. We assume that in practice, archival of a set of data into a preservation system, is done at critical points in the product lifecycle. Such dedicated points in time are often the transitions between different phases of the product lifecycle. Typical points would be the transition between the ideation phase and the development phase, or the transition between development and production of a product. The phase before the archival can be arbitrary long. Thus, it is difficult to manually recreate the knowledge about social and collaboration context at archival time for reasons of human memory, availability, and cost. This paper discusses an approach to enable the tools used during the creation of the product data to become aware of the social and collaboration context, to enable them to express themselves in the terms of these contexts, and to capture the knowledge for preservation and reuse semi-automatically. The paper is structured as follows. First the requirements for making applications preservation-aware are discussed, and then related work is presented. The next sections discuss the OpenConjurer approach of providing a flexible ontology-based vocabulary and infrastructure. This is followed by the introduction of a demonstration application which is used to illustrate how to implement a preservation aware collaborative application in practice. Then it is discussed how to integrate the preservation aware applications with content repositories, product lifecycle management (PLM) systems, and preservation systems. The paper concludes with a discussion and an outlook to future work.

2 Requirements

The problem addressed by this paper is the question on how to enable the tools used during the creation of the product data to become aware of the social and collaboration context, to enable them to express themselves in the terms of these contexts, and to capture the knowledge for preservation and reuse semi-automatically. The first sub-problem in order to capture such processes and social context is to specify the matching vocabulary describing the social context as well as the collaboration context. As individual enterprises differ significantly regarding their structure and corporate culture, the vocabulary needs to be customizable to match the real world context of use. In addition to the vocabulary it is also necessary to operationalize the management of an inter-organizational social graph by providing an infrastructure and APIs. In practice, product data is stored across a very heterogeneous population of document management systems, file shares, and PLM systems. In order to enable the capturing of contextual metadata, product data repositories are necessary that offer interfaces for annotating product data with custom metadata. For integration with the overall workflows such as decision making or archival the repositories need to be linked to legacy PLM systems. For archival, the repositories need to provide interfaces for collecting the product data and collected context metadata.

3 Related Work

Gathering information about what happened in a collaborative environment is not a new concept, and is called activity log, event history, or elephant's brain. From a software engineering perspective this can be considered a design pattern and is described in detail by Schümmer and Lukosch, 2007 [14]. The intent of this pattern is to: "Store information about users' activities in a log to provide a history of their activities and the artifacts' evolution." This pattern is regularly applied in various collaboration environments, and is especially prominent in software development environments. A system keeping an activity log is already capturing a basic social context (user names) and collaboration context (what has been done). However, usually the type of activities to be logged is hard wired, and the vocabulary is not customizable.

Models for social graphs already exist. Two of the most popular vocabularies are the friend of a friend (FOAF) [6] ontology and Xhtml Friends Network (XFN) [17]. FOAF models a distributed social network, by offering an RDF vocabulary to be embedded into web pages or other documents. The vocabulary already includes specific concepts and properties, which do not apply to all use-case scenarios.

XFN is a micro format allowing to add machine interpretable markup into HTML documents, modeling a social graph. A notable concept provided by XFN is the possibility to consolidate identities by so-called me-links. A me-link expresses that two online identities are essentially the same person.

Fig. 1 The Social Ontologies
Hierarchy

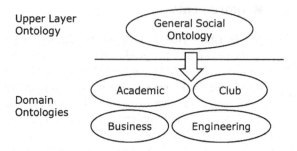

In digital preservation, the OAIS reference model [5] is commonly adopted and illustrates the common understanding of preservation systems and workflows adopted by the digital preservation community. In OAIS, the tools used for content creation are mostly unaware of the archival system. Data from a producer is put into a so-called Submission Information Package (SIP) which is ingested into the archival system and is transformed into a so-called Archival Information Package (AIP) and is preserved within the archive. In essence, this means that the actual data from the producer is created independently from the preservation environment. Usually during the creation of the SIP, metadata for the data is cleaned up and added, either automatically or with manual intervention. The result of this view of preservation being completely decoupled from the actual data creation results in scenarios, where metadata which is readily available during data creation is lost before reaching the creation of the AIP. For example, change logs and user information are available in MS Office documents, but in practice digital libraries often only archive a PDF version of the source material. In the conversion process usually most of the metadata is lost. While it is certainly possible to archive the original documents, the loose coupling of the preservation systems with the content creation environment and the lack of awareness of the content creation tools for requirements from preservation result in this common loss of important data.

This paper proposes to make the tools and environments used during data or content creation more preservation-aware by providing a common framework for expressing social and collaboration context.

4 Vocabulary

The organizational context of the usage of a collaboration environment in design and engineering is not a priori well defined. The reason for this is that in each organization or enterprise the corporate culture may vary significantly. For example some organizations use a very hierarchical organization, others are structured very flat. In addition, the vocabulary in which a hierarchy and relations between individuals are expressed also varies from organization to organization. For example, in a university a relation to express hierarchy may be "X is PhD advisor of Y", while in a

Fig. 2 The General Social
Ontology

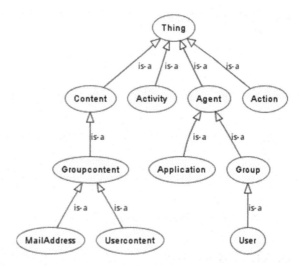

design and engineering company "X is supervisor of Y" may express a relation
with the same operational consequences, but with a different vocabulary. Thus, a
customizable approach is required in order to ensure the adaptability of the resulting
system to changing target environments. In order to accommodate the wide range of
possibilities for organizational structures, this paper proposes to use a minimal upper
layer ontology to express the basic concepts of social and collaboration context. This
upper layer ontology is called general social ontology (GSO) and is supposed to be
adapted to practical scenarios by specifying domain ontologies extending the GSO,
as illustrated in Figure 1.

The description of the whole ontology and approach is beyond the scope of this
paper. Figure 2 illustrates the class hierarchy of the ontology. Active entities in the
system are modeled as agents. Agents can be applications, groups, and users. Users
are a sub-class of Group, as a single user can have multiple identities and the union
(grouping) of all identities of a user can be considered to identify a single user.
Profile data is modeled under Content. In order to describe collaboration context,
Action and "Activity" are provided. "Action" provides a concept for describing the
type of something that can be done within a collaborative environment, or with
data. Activities are then the actual instances of Action which are actually performed.
In this paper, a log of activities is exactly what provides collaboration metadata.
Further information making up a social graph, such as relations and details, are
modeled as object or data properties.

Figure 3 illustrates how, for example, the data in the profiles of users can be
customized by extending the User Content concept. In a similar way, individual
relations can be specified.

Fig. 3 A Domain Ontology extending parts of the General Social Ontology

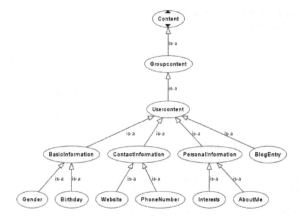

Fig. 4 The OpenConjurer Infrastructure

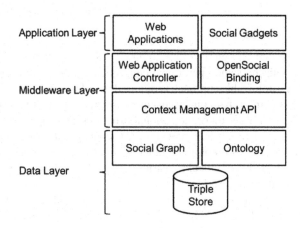

5   Infrastructure and API

The framework OpenConjurer realizes the concept of a social collaborative platform in order to enable the actual tools used for content creation, design, engineering, coordination and collaboration to become aware of social context, collaboration context, and preservation.

As a first step it is necessary to provide an identity management infrastructure (e.g., see Maler and Reed 2008 [12]) for managing the identities of users, groups, and relations.

Figure 4 illustrates the tiered approach of the OpenConjurer infrastructure. On the data layer, the social graph is expressed by means of instances of the vocabulary as discussed in the previous section. The social graph consists of the general social ontology, a domain ontology, and the triple data describing the actual individuals, the user data, and relations. This data is persisted in a triple-store. In the prototypical implementation, the data layer is implemented using the Jena [10] sematic

Fig. 5 The OpenConjurer
binding to OpenSocial

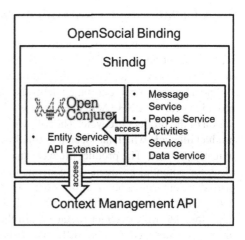

web framework, offering SPARQL [16] queries on the social graph and persistent storage.

The middleware layer of OpenConjurer is based on the context management API which is designed to abstract from the basic SPARQL queries used to interact with the data tier. Instead the context management API provides an interface which is much more natural for the actual interaction with a social graph.

On top of the context management API, OpenConjurer implements a container for social web applications, similar to social networking sites like Facebook and Orkut. In addition, OpenConjurer implements an OpenSocial [7] container, i.e., a social web application that can execute custom social applications.

OpenConjurer uses Apache Shindig [2], a Java EE reference implementation of OpenSocial. This reference implementation offers the possibility to expose any social network via OpenSocial APIs. However, in order to bind these APIs to the data layer of OpenConjurer, an appropriate bridging mechanism has to be implemented. Figure 5 illustrates the OpenSocial binding to OpenConjurer. OpenSocial itself is not able to express itself through the same kind of configurable vocabulary as the general social ontology bundled with specific domain ontologies.

In order to resolve these problems, i.e., bridging and limited semantic expressivity of OpenSocial, OpenConjurer implements an API extension to OpenSocial, the so-called entity service. The entity service offers calls to access the context API using the OpenConjurer ontologies. In this way, OpenSocial applications harness the expressiveness of the OpenConjurer models. To provide the bridging between the stub implementations of the OpenSocial services provided by Apache Shindig and OpenConjurer, OpenConjurer implements a mapping of the individual service calls to the matching calls of the entity service. This way the separation of concerns within OpenConjurer is maximized, as the context management API can be kept clean, only dealing with OpenConjurer concepts, without directly emulating OpenSocial, and the implementation of the mapping can be limited to the actual OpenSocial implementation.

6 A Collaborative Preservation-Aware Application

This section describes a collaborative application and how exactly such a collaborative application is turned into an application which is aware of the social context, the collaboration context, and preservation by using the OpenConjurer framework.

The application is based on an analysis of the design and engineering scenarios provided within the SHAMAN research project. These scenarios are built around the notion of the product lifecycle. The following is a simplified description of a typical product lifecycle which has been identified within practice of the SHAMAN project industry partners. The product lifecycle starts with the ideation process, where new ideas for potential products are generated and evaluated. Next, ideas that are approved are entering a phase of prototyping, where first designs are created and evaluated. Once prototyping is successful and after approval of the project, the actual development of the product is starting. In the consumer electronics industries that have been examined, a major difference between the prototyping and development phase is that the prototyping is considered to be a very creative process. Thus, the companies do not constrict their developers by specifying which tools shall be used. The developers have free choice of tools during this time in order to maximize creative freedom, resulting in a very heterogeneous environment. Once entering the actual product development, the tools to be used are pre-defined very specifically in order to provide an efficient process and quick time to market. Following the development, the product lifecycle enters the production phase, followed by maintenance and finally the disposal and recycling of the product.

The prototype application described here is residing in the prototyping or development phase. In practice, companies are reluctant to provide access to the actual live development environments, tools, and data. The reasons for this are very clear. The company would risk leaking of confidential intellectual property, as well as disruption of critical processes. In addition, in practical environments specialized development tools are used which are very expensive, customized, and have very specific licensing schemes making it hard to customize within the context of a research project.

Thus, an example collaborative application, based on the product lifecycle has been developed as a proof of concept demonstrator. The demo application, the so-called decision room, emulates a synchronous meeting for reviewing product data and for managing related tasks, e.g., for deciding if a product shall make the transition from prototyping to the design phase. The tool itself is not designed with the intention of providing domain specific support for decision-making, but to contain key components of collaborative software to illustrate the capturing of social and collaboration context during content creations and to demonstrate some new ideas of how to dynamically scope collaboration based on documents in a repository.

The decision room application is a social web application. The implementation is based on the OpenConjurer OpenSocial implementation as a social Gadget, using the OpenSocial extensions to the API. The application consists of several collaboration widgets, as illustrated by Figure 6:

Fig. 6 The Decision Room Application

a) Collaborative content repository browser
b) Group task list
c) Document centric chat
d) Collaborative viewer

The individual widgets are described in the following sections.

6.1 The Repository Browser

The concept of the decision room is based on a document centric scoping mechanism for collaboration. The repository browser is used to select the current scope,
i.e., a folder or document of the user. The scope is signaled to all other widgets
which update their view accordingly.

The repository browser is the front-end to an actual content repository based on
the Java Content Repository [11] specification, using the Apache Jackrabbit implementation [1]. This hierarchical content repository provides extended capabilities,
by explicitly managing metadata for each individual folder and document stored
inside. For each document or folder an additional document of metadata can be associated. These metadata documents are not visible while browsing the repository,
but can be accessed directly if required via an API.

The collaborative repository browser provides a tree view of the data repository.
The users can browse the structure and content of the repository. The scope of a user
for the entire decision room is set by selecting a node in the tree view representing

either a folder, or a document in the repository. The collaborative component of the repository browser is the support for document centric collaboration awareness. In the tree view of the repository browser, every node is annotated with the number of users, and the respective user names, which have selected the node as their individual scope. In this way, the users know where within a project or product their team members are working and can join them for collaboration, using the other collaboration widgets.

6.2   The Group Task List

The group task list collaboration widget allows the users to assign tasks to individual scopes (documents or projects). Users can create tasks, edit them, and mark them as done. The group task list is a very simple tool, which is used to illustrate the capturing of process-oriented collaboration metadata, which would be available from workflow for decision support systems.

6.3   Document Centric Chat

For demonstrating the capturing of contextualized communication the document centric chat allows users to exchange text messages in real time. For the chat the scoping of the decision room on a single document or file acts similar to the concept of chat rooms or chat channels. The chat widget automatically provides an individual chat room for each folder and document. Collaborative Viewer

To access the actual design data, the collaborative viewer widget provides a shared viewing environment. Once the scope is set to a document, the viewer loads the documents and renders a view. The viewer provides a synchronized view for all users who have selected the same document at the same time as their individual focus in the repository browser. The viewer supports zooming and panning. These operations are also replicated for all users with the document as focus, resulting in an identical view for all users.

Currently, the collaborative viewer is limited to bitmap graphics. However, development of integration with multivalent viewing technology [13] is in progress, enabling the viewing of different formats such as JT [15] for 3D CAD models, which stems from Siemens PLM technology and is widely adopted as a visualization and exchange format in industry.

7   Making the Application Preservation-Aware

By implementing the decision room on top of the OpenConjurer framework the individual widgets can access the extended OpenSocial API, and can express themselves regarding the social context. The widgets can identify the user, the repository, scope, and the activities performed by the users. Using the API provided by the content repository, the decision room can add such captured metadata to the individual documents in the repository.

7.1 Actions and Activities

As discussed above, the GSO differentiates between actions and activities. An action determines the type of something that can be done with an application in a certain scope. For example, within the collaborative repository browser typical actions are to select and unselect a scope. An activity is the actual event or process of performing an action, such as selecting a specific scope in the repository browser, or taking the decision that a design is production ready. An action consists of:

- the scope of the action: a unique identifier pointing to the repository, folder and document,
- the applications involved in performing the action: a unique identification of the application, version, and a description of the runtime environment,
- the users involved in the action: user IDs from the social graph,
- the type of the action: unique identifier of the action,
- a timestamp: a timestamp when the action occurred. The time (client/server) to use depends on the action.
- Action specific information: Actions can be very domain specific. Thus, applications need to be able to add domain and action specific information. For the decision room, this can be any XML encoded content.

7.2 Capturing Social Context and Collaboration Context

As OpenConjurer is designed as a generic framework for capturing social and collaboration context. Thus, the framework is not informed in advance about which tools will be built on top of it, and it is not possible to identify the relevant actions in advance. In addition, the type of actions is domain specific within the individual tools for content creation, design and engineering tools.

In order to make an application preservation-aware, domain experts designing the application have to identify the relevant actions to be captured. Typical generic actions are: read, change, delete, and create. However, more specific action such as triggering a computation intensive simulation process on a new electronic design may be relevant and are directly depending on the type of application at hand. Table 1 lists the actions that were identified as relevant within the decision room application.

In order to be able to capture the activity log, the application must know the following parameters:

- scope of an activity to be performed
- users using the application and triggering an activity

In the case of the decision room, the scope is determined through the repository browser by selecting a folder or document. In addition, the decision room

Table 1 Example activities to be tracked in the collaborative application

Application	Action Type	Scope	Users	Timestamp	Action specific information
Repository Browser	Select Scope	scope selected by user	user selecting the scope	Server	NA
Repository Browser	Unselect Scope	scope unselected by user	user unselecting the scope	Server	NA
Task List	Add Task	current scope of user adding the task	user adding the task	Server	Task Description
Task List	Delete Task	current scope of user deleting the task	user deleting the task	Server	Task Description
Task List	Edit Task	current scope of user editing the task	user deleting the task	Server	Resulting Task Description
Task List	Mark Done	current scope of user marking the task	user marking the task	Server	Task Description
Task List	Mark Not Done	current scope of user marking the task	user marking the task	Server	Task Description
Chat	Chat Message	current scope of user posting the chat message	user posting the chat message	Server	Chat Message
Viewer	Configure View	current scope of user changing view	user changing view	Server	Pan and Zoom

is implemented as an OpenSocial application running inside a custom container. Thus, the application can easily identify the user through the appropriate extended OpenSocial API calls. In general, the scope of an action is again determined in a very domain specific way which cannot be predicted on the framework level. However, the knowledge about the users' identity can be gained by using the OpenConjurer framework as an identity management back end. In addition to the OpenSocial API, OpenConjurer also offers web services providing the same information to other applications.

Once the list of actions has been identified the domain experts have to identify the appropriate parts within the applications code, where these actions are actually performed. At these points within the code, the individual activities have to be captured and the metadata set for each activity has to be written into the metadata set of the current scope. In the decision room application, the used content repository offers the matching scoped web services to collect such activities. In this case this means that XML fragments describing activities are collected for each folder and document.

Once this metadata is available in the working repository or PLM systems, it can be taken account of during an archival process and ingest into an archive. An application actively collecting such social and collaboration metadata during content creation time is called preservation aware. The next section describes how such a preservation-aware system can be integrated with a PLM system and the SHAMAN preservation environment.

7.3   Integration with PLM and Preservation Environment

In interviews with industry, it became obvious that in practice a company does often not rely on a single content repository, document management, or product lifecycle system. Product data is often distributed across a heterogeneous population of systems, and needs to be consolidated and normalized for preservation.

As discussed before, access to live production systems was not available. Thus, we simulate such a real-world scenario, by providing two different repositories. On the one hand we have a professional product lifecycle management system, i.e., Aras Innovator [3], and on the other hand we have the JCR-based repository described above. In order to demonstrate integration and data harmonization, we defined the Aras PLM system as the so-called leading system which is used to trigger the preservation process. Within the PLM system, the JCR repository of the decision room is referenced. This reference allows access to the content of the repository.

Figure 7 illustrates the integration of the decision room and the product lifecycle management for preservation. On the left hand side, the decision room application runs on top of the OpenConjurer framework, using a demo domain ontology which is an extension of the general social ontology. Within the decision room, users work on documents stored within the content repository.

When the preservation is triggered from within the PLM system, the preservation component ADI traverses the content in the PLM system, building a normalized information package for the ingest process of the SHAMAN preservation environment.

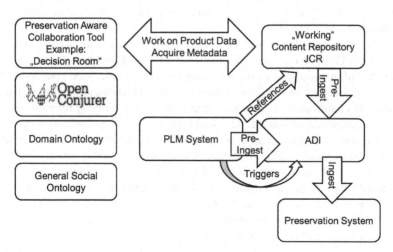

Fig. 7 Integration with preservation system

When the ADI traverses a node within the PLM containing a link to a different content repository, the ADI continues by traversing the target content repository and integrates into the information package for ingest. The union of the content of both systems is harmonized on a data level before the actual ingest is done. The actual normalization process is out of scope of this paper and will be documented in future papers.

8 Results and Discussion

The problem addressed by this paper is the question on how to enable the tools used during the creation of the product data to become aware of the social and collaboration context, to enable them to express themselves in the terms of these contexts, and to capture the knowledge for preservation and reuse semi automatically. The general social ontology presented in this paper provides the base for a vocabulary describing the social context as well as the collaboration context. The differences between individual enterprises can be addressed by domain ontologies extending and customizing the general social ontology. In addition to the vocabulary, it is also necessary to operationalize the management of an inter-organizational social graph by providing an infrastructure and APIs. This is realized through the OpenConjurer framework, providing access to the social graph via web services and an extended OpenSocial binding. However, currently OpenConjurer is not able to handle cross-organizational or inter-enterprise scenarios. As discussed above, product data is stored across a heterogeneous population of document management systems, file shares, and PLM systems. An example of a repository that supports the capturing of contextual metadata, the scoped metadata collection in the extended JCR repository has been presented above. The integration with the overall workflows such as decision making or archival the repository has been integrated with a legacy PLM system. This paper also outlined the basic prerequisites for making applications preservation-aware and the basic steps in the implementation of preservation-aware collaborative applications.

The work presented here has been evaluated in a focus group meeting with six experts from Philips Consumer Lifestyle. In this focus group, the software has been demonstrated, a questionaire, and a plenary discussion have been conducted. The results are currently under evaluation. However, the experiences from the event point towards the confirmation, that capturing metadata early is highly relevant and that the collaboration features based on a social network would be of interest to the users. Thus, we conclude that the approach to gather metadata during content creation through a facilitating social collaboration middleware is valid. During the focus group meeting it became apperant, that due to the short lived nature of the products designed by the selected target audience, the long term prevervation aspects where not as critical. It was suggested to investigate with an audience from the departments of medical products instead of consumer electronics, as the requirements regarding preservation are more well defined for such products.

9 Conclusion and Future Work

In this paper we have illustrated an approach to make the data and content creation tools aware of social and collaboration context as well as aware of preservation by providing vocabulary, frameworks and APIs that enable the applications to express themselves in terms of context, to access context information, and to annotate data with captured domain specific metadata. This way, new categories of metadata become available to preservation systems, which previously were not available because of cost and complexity of generating the metadata at ingest time. The presence of this metadata supports the users' needs for knowledge reuse regarding the processes within the product lifecycle.

This paper only looked at products developed within a single organization. However, in practice products have reached a complexity that makes it impossible for a single enterprise to create a product on their own. Research and development as well as production now often happen within a complex network of enterprises, resulting in the emerging rise of virtual enterprises and new more service-oriented business models for manufacturing industries. It will be necessary to extend the models and frameworks presented in this paper to accommodate such inter-enterprise distributed scenarios, resulting in various interoperability problems. One of these problems is the question on how to express relations between two organizations which are using different vocabularies for modeling their social graph.

By providing the notion of a customizable social graph, expressing the corporate culture and relations between individuals, projects, and data, it may also be feasible to use the social graph of an enterprise, or even a network of organizations for setting of a federated identity management environment which can be integrated with access rights management.

Acknowledgements. This work has been partially supported by the FP7 EU Large-scale Integrating Project SHAMAN (Sustaining Heritage Access through Multivalent ArchiviNg), co-financed by the European Union. For more details, visit `http://shaman-ip.eu/shaman/`

References

1. Apache Software Foundation: Apache Jackrabbit, `http://jackrabbit.apache.org` (cited March 20, 2011)
2. Apache Software Foundation: Apache Shindig, `http://shindig.apache.org/`
3. Aras Corp: Aras PLM Software, `http://www.aras.com` (cited March 20, 2011)
4. Brickley, D., Miller, L.: FOAF Vocabulary Specification 0.98. Namespace Document, Marco Polo Edition (August 9, 2010), `http://xmlns.com/foaf/spec/` (cited March 20, 2011)
5. Consultative Committee for Space Data Systems: Reference Model for an Open Archival Information System (OAIS), CCSDS 650.0-B-1, Blue Book (2002)
6. FOAF: The Friend of a Friend (FOAF) project, `http://www.foaf-project.org` (cited March 20, 2011)
7. Google: Google Code OpenSocial, `http://code.google.com/intl/en/apis/opensocial` (cited March 20, 2011)

8. Heutelbeck, D., Brunsmann, J., Wilkes, W., Hemmje, M., Hundsdörfer, A., Heidbrink, H.-U.: Towards support for long-term digital preservation in product life cycle management. In: Proceedings of the 6th International Conference on Preservation of Digital Objects, iPRES 2009, San Francisco, United States, October 5-6, pp. 211–219. California Digital Library, UC Office of the President (2009)

9. Heutelbeck, D., Brunsmann, J., Wilkes, W., Hundsdörfer, A.: Motivations and Challenges for Digital Preservation in Design and Engineering. In: InDP 2009, First International Workshop on Innovation in Digital Preservation in Conjunction with JCDL 2009, Austin TX (2009)

10. Labs, H.P.: Jena Semantic Web Framework, http://jena.sourceforge.net (cited March 20, 2011)

11. Java Specification Request: JSR-170 – Content Repository for JavaTM technology API, http://www.jcp.org/en/jsr/detail?id=170 (cited March 20, 2011)

12. Maler, E., Reed, D.: The Venn of Identity – Options and Issues in Federated Identity Management. IEEE Security & Privacy, 16–23 (2008)

13. Phelps, T.A., Watry, P.B.: A No-Compromises Architecture for Digital Document Preservation. In: Rauber, A., Christodoulakis, S., Tjoa, A.M. (eds.) ECDL 2005. LNCS, vol. 3652, pp. 266–277. Springer, Heidelberg (2005)

14. Schümmer, T., Lukosch, S.: Patterns for computer mediated interaction. John Wiley & Sons Ltd. (2007)

15. Siemens: JT Open, http://www.plm.automation.siemens.com/en_us/products/open/jtopen/ (cited March 20, 2011)

16. W3C: SPARQL Query Language for RDF, http://www.w3.org/TR/rdf-sparql-query (cited March 20, 2011)

17. XFN: Xhtml Friends Network, http://gmpg.org/xfn (cited March 20, 2011)

Univector Field Method Based Multi-robot Navigation for Pursuit Problem

Hoang Huu Viet, Sang Hyeok An, and TaeChoong Chung

Abstract. This paper introduces a new approach to solve the pursuit problem based on a univector field method. In our proposed method, a set of robots work together instantaneously to find suitable moving directions and follow the univector field to surround and capture a prey robot. In addition, a set of strategies is proposed to make the pursuit problem more realistic in the real world environment. This is a general approach based on univector field, and it can be extended for an environment that contains static or moving obstacles. Experimental results show that our proposed algorithm is effective for the pursuit problem.

Keywords: univector field, predator robots, prey robot, pursuit problem.

1 Introduction

The pursuit problem is a well-known class of test problems for the study of cooperative behavior in Distributed Artificial Intelligence (DAI) systems. This problem was first introduced by Benda et al. [1], and has recently been received a great deal of attention from the research community due to its important role in a wide range of applications, especially in some areas such as robotics and computer games. In Benda's formulation, a group of four predator agents try to capture a prey agent by surrounding it from four directions on a grid-world. Agent movements are limited to either a horizontal or a vertical step per time unit. The prey agent moves randomly, and two predator agents are allowed to occupy the same location. He proposed a solution to this problem by using computations of the center of gravity of the agents.

Gasser et al. [3], introduces an approach to this problem in which the grid-world is divided into four quadrants by using diagonal lines crossing at the prey agent's

Hoang Huu Viet · Sang Hyeok An · TaeChoong Chung
Artificial Intelligence Lab, Department of Computer Engineering,
School of Electronics and Information, Kyung Hee University,
1-Seocheon, Giheung, Yongin, Gyeonggi, 446-701, South Korea
e-mail: {viethh,ash,tcchung}@khu.ac.kr

J. Altmann et al. (Eds.): Advances in Collective Intelligence 2011, AISC 113, pp. 131–143.
springerlink.com   © Springer-Verlag Berlin Heidelberg 2012

position. Each of the predator agents tries to occupy a position within their assigned quadrant to attain the "Lieb configuration". Once the predator agents have attained the "Lieb configuration", they use a set of rules called the "Lieb rules" that determine the movements of the predators to capture the prey agent. However, no experimental results were presented for the performance of this approach.

Stephens and Merx [8] present a version of the pursuit problem that does not allow two agents to occupy the same position. The predators alternated moves and the prey moves randomly. In their approach, a centralized control mechanism is used to implement a scheme similar to the Lieb rules of [3] for the predator agents.

Korf [6] proposes a greedy algorithm to solve the pursuit problem. In his approach, each predator agent measures the distances between coordinates of its empty neighboring cells and the prey agent's coordinates, and then moves to a cell that minimizes distance to the prey agent's position. Three versions of the pursuit problem are presented. The first version is called the orthogonal game that corresponds to the original version, where agent movements are horizontal or vertical. He claims that a discretization of the continuous world with only horizontal and vertical movements is poor approximation. Therefore, he develops the diagonal game that allows eight agents to move orthogonally, diagonally, and then surrounds the prey. Finally, the hexagonal game is implemented on an environment of hexagonal grid, where each cell has six neighboring cells and there are six predator agents. In the first and second versions, a max norm distance metric is used by the predator agents to choose their next steps. However, the Euclidean distance metric is used instead of the max norm distance metric in the third version. He also improves the performance of the algorithm by adding an "attractive forces" between the predators and the prey, and "repulsive forces" between the predators. He assumes that the predators have the knowledge of existence of all the other predators and the prey. As a result, every single predator computes the "resultant force" to choose their next move.

In Korf's model, agents are able to see the complete world at once. Therefore, Juan Reverte et. al. [7], extend Korf's ideas to improve his model. First, they propose a simple extension of Korf's fitness function and consider the problems related to a partial view of the world. Second, they propose a communication protocol to partially overcome them. They conclude that with these two simple extensions, Korf's ideas get results comparable to most machine learning approaches.

Although all the approaches mentioned above might potentially solve the pursuit problem. However, in our opinion, these approaches are insufficient to implement in the real world environment for mobile robots. They are only applicable for the simulated pursuit games. Because when robots are used instead of agents in the pursuit problem, the posture of the robots at each position of the environment need be considered. Therefore, we propose a new approach for the pursuit problem that can be applied to real world environment in the area of robotics. In our approach, a set of predator agents, called predator robots, work together to find suitable moving directions and follow the univector field of the environment to surround and capture a prey agent, called prey robot. In addition, strategies of the predator robots and prey robot are defined to test the proposed approach. The experimental results prove that the proposed approach guarantees to find a solution of the pursuit problem in the finite moving steps of the predator robots if the environment is large enough. Moreover, our approach provides a general solution since with minor modifications it can be applied to multi-agent real-time pursuit (MAPS) problem in a partially observable environment with obstacles.

The rest of this paper is organized as follows: Section 2 overviews related background knowledge consisting of the two-wheel robot modeling and the univector field method. The proposed approach is described in Section 3. Section 4 shows the experimental results. Finally, we conclude our work in Section 5.

2 Background

2.1 Two-Wheel Robot Modeling

The mechanical structure of a two-wheel robot is shown in Fig. 1 [2, 5], where L is the base width of the robot and R is the radius of the wheel. It is assumed that the position and orientation of the robot are known at each instant time through sensory inputs. Let ω_L and ω_R be the rotational velocities of the left and the right wheels respectively, and assume that no slipping of the wheels, the wheel velocities at the contact points are determined as in equation (1),

$$V_L = R\omega_L, \quad V_R = R\omega_R \tag{1}$$

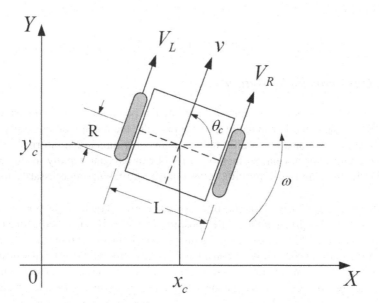

Fig. 1 Two-wheel robot modeling

Let v be the velocity of the robot at its center and ω be the turning velocity with respect to the center of the robot. The velocity vector s is defined as in equation (2),

$$s = \begin{bmatrix} v \\ \omega \end{bmatrix} = \begin{bmatrix} \frac{V_R+V_L}{2} \\ \frac{V_R-V_L}{L} \end{bmatrix} = \begin{bmatrix} \frac{1}{2} & \frac{1}{2} \\ \frac{-1}{L} & \frac{1}{L} \end{bmatrix} \begin{bmatrix} V_L \\ V_R \end{bmatrix} \qquad (2)$$

The kinematics of the robot is illustrated in Fig. 1. Posture p_s and position p of the robot are defined as in equation (3), where (x_c, y_c) is position of the center of the robot, and θ_c is the heading angle of the robot with respect to absolute coordinates (x_c, y_c).

$$p_s = \begin{bmatrix} x_c \\ y_c \\ \theta_c \end{bmatrix}, \quad p = \begin{bmatrix} x_c \\ y_c \end{bmatrix} \qquad (3)$$

The robot kinematics associated with the Jacobian matrix and velocity vector s is defined as in equation (4),

$$\dot{p}_s = \begin{bmatrix} \dot{x}_c \\ \dot{y}_c \\ \dot{\theta}_c \end{bmatrix} = \begin{bmatrix} \cos\theta_c & 0 \\ \sin\theta_c & 0 \\ 0 & 1 \end{bmatrix} \begin{bmatrix} v \\ \omega \end{bmatrix} \qquad (4)$$

To get the robot position and orientation, equation (4) should satisfy the nonholonomic constraint as shown in equation (5), which is equivalent to $\frac{dy_c}{dx_c} = \tan\theta_c$. This means that the moving direction at every instant time is the same as the heading angle of robots.

$$\dot{x}_c \sin\theta_c - \dot{y}_c \cos\theta_c = 0 \qquad (5)$$

2.2 Univector Field Method

Univector field method is an improvement of the potential vector field which is designed for fast mobile robots. In a univector field, the magnitude of each vector is unity at all positions, so a vector field contains only directional information. By using the univector navigation method, a robot can navigate rapidly to the target position with desired posture without oscillating and taking unnecessarily longer paths [5].

The univector field is classified into two categories. The first category is concerned with a robot going at the desired posture. The other is concerned with a robot avoiding obstacles. Since we only solve the pursuit problem in an obstacle-free environment, so the first category is considered. A univector field $F(p)$ at the position $p(x, y)$ is defined as in equation (6), where n is a positive constant to be determined, g is the desired goal point, r is a guidance point, and the symbol $\angle$ denotes the angle of a vector mapped onto the range $(-\pi, \pi]$. The value of the univector field $F(p)$ at the position $p(x,y)$ is equivalent to the desired heading angle θ_c of the robot as in equation (3).

$$\angle F(p) = \angle \overline{pg} - n\alpha \quad \text{with} \quad \alpha = \angle \overline{pr} - \overline{pg} \tag{6}$$

All univector fields constitute a univector field space $N(p)$ for an environment of the robot. Fig. 2 shows a univector field space, where each tiny circle represents a position and the straight line attached to it represents a moving direction.

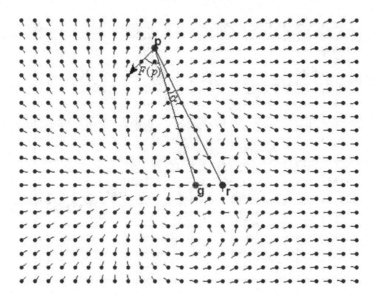

Fig. 2 Univector field space

It is clear that the larger n is, the smaller the $F(p)$ is at the same robot position. Thus, if n increases, the univector field spreads out to a larger area, making the path to be traversed by the robot in reaching its goal larger. The univector field $F(p)$ of the robot at the position $p(x,y)$ changes according to the parameter n and the length of the line gr.

3   The Pursuit Problem

In the classical pursuit games [1], a group of four predator agents try to capture a prey agent by surrounding it from four directions on a grid-world. In this paper, the model of two-wheel robots is used for pursuit problem, called predator robots and prey robot, instead of the predator agents and prey agent respectively. To solve this problem, a Univector Field method based Multi-Robot Navigation (UFMRN) is proposed to make the pursuit problem more realistic in the real world environment.

Our pursuit problem is described as follows: It is assumed that there are eight predator robots and a prey robot within the system. The simulations allow all robots to move to any of eight neighboring cells, including horizontal move, vertical move and diagonal moves. The solution to the pursuit problem is to completely surround a prey robot with eight predator robots as shown in Figure 3. In the next subsections, we are going to propose strategies of robots and an UFMRN algorithm to solve this problem.

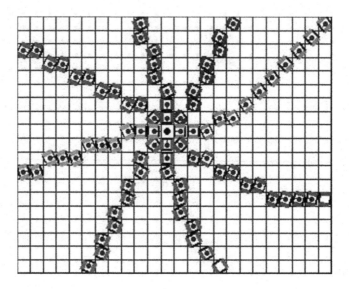

Fig. 3 A solution to the pursuit problem based on univector field, the circle is the prey robot

3.1 Strategies of Predator Robots

a) Greedy strategy

It is assumed that each predator robot knows only the position of the prey robot at each instant time through sensory inputs. Each predator robot measures the distances from its position to empty neighboring cells of the prey robot's position, and moves to an empty neighboring cell that has the minimum distance.

Fig. 4(a) depicts a situation of the greedy strategy in which the predator robots 1, 2, 3, 5 and 7 chose a neighboring cell of the prey robot for each of them that minimizes distance to the neighboring cell of the prey robot. Therefore, the predator robots 4, 6 and 8 have to choose a cell from the empty neighboring cells that has the minimum distance from its position to those cells.

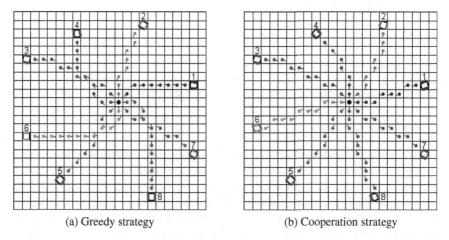

(a) Greedy strategy   (b) Cooperation strategy

Fig. 4 The capture directions of the predator robots based on greedy and cooperation strategy

b) Cooperation strategy

The cooperation is one of the most important strategies in making an agent community more intelligent to achieve a common goal. The goal of the cooperation strategy is to share short-term benefits of agents to increase efficiency of overall agents for a task. Accordingly, we assume that each predator robot communicates to each other to know positions and orientations of the other predator robots and the prey robot at each instant time through sensory inputs. The predator robots cooperate to determine a strategy that has the minimum sum of distances from all predator robots to eight neighboring cells of the prey robot. Fig. 4(b) shows a situation that the predator robots occupy positions as the Fig. 4(a), but by using cooperation strategy, the predator robots share directions to surround the prey robot.

From Fig. 4, it is clear that by applying the univector field method for the pursuit problem, the directions of robots 1, 4, 6 and 8 change when the capture direction of these robots changes.

c) Mixed strategy

To make the pursuit problem more realistic, we consider the pursuit problem related to a partial observation of the environment and propose a mixed strategy of greedy strategy and cooperation strategy. It is assumed that each predator robot knows the position of prey robot at each instant time, but it only communicates its position and direction with some predator robots in a predefined radius. If a predator robot cannot communicate its information to the other predator robots, a greedy strategy is applied for this predator robot. Otherwise, a cooperation strategy is applied for predator robots in such a manner that they can communicate to each other.

3.2 Strategies of Prey Robot

a) Random strategy

It is assumed that prey robot does not know about positions and directions of the predator robots. At each instant time, the prey robot moves randomly to an empty neighboring cell from the set of its eight neighboring cells.

b) Greedy strategy

It is assumed that the prey robot knows about positions and directions of the predator robots at each instant time. In this strategy, the prey robot measures the distances to eight predator robots and moves an empty neighboring cell from the set of its eight neighboring cells that maximizes the distance to the nearest predator robot. However, if each predator robot moves at the same speed as the prey robot and the positions of eight predator robots are initialized at the same side of the prey position, then the predator robots cannot capture the prey robot. Therefore, we assume that the prey robot will move with a probability of $q_0(0 < q > 0 < 1)$ and it will remain stationary with a probability of $1 - q_0$.

c) ε-greedy strategy

The combination of the random strategy and the greedy strategy makes an ε-greedy strategy. That is, with small probability $\varepsilon(0 < \varepsilon < 1)$, the random strategy is applied for the prey robot, and with probability $1 - \varepsilon$, the greedy strategy is applied for the prey robot.

3.3 The Proposed Algorithm

An UFMRN algorithm is proposed for the pursuit problem as in Algorithm 1. This algorithm implements the mixed strategy of the predator robots and the ε-greedy strategy of the prey robot. With minor modifications, the other strategies can be easily applied to this algorithm. Parameters of UFMRN algorithm are described as follows:

$p(x,y)$ is the position at the coordination (x,y) of the environment.
$A(k,p)$ is the current position of the k-th predator robot $(k = 1,\ldots,8)$.
$F(k,p)$ is a value of the univector field at $p(x,y)$ of the k-th predator robot.
$g(x,y)$ is the position at the coordination (x,y) of the prey robot.
$D(k,g)$ is a guidance point of the k-th predator robot (it's a neighboring cell of $g(x,y)$).
$C(k,g)$ is a neighboring cell of the prey robot's position for the k-th predator robot.
r is a radius parameter of the mixed strategy.
ε is a parameter of the ε-greedy strategy.
q_0 is a parameter of the greedy strategy of the prey robot.

Algorithm 1. The UFMRN algorithm for the pursuit problem

1: Initialize $A(k,p)$ arbitrarily for all $k = 1, \ldots, 8$
2: Initialize $g(x,y)$ arbitrarily
3: $stop = false$
4: $t = 0$
5: **repeat**
6:     **for** $k = 1$ **to** 8 **do**
7:         $C(k,g) = 0$
8:         **if** the k-th predator robot does not find a predator robot in the radius r **then**
9:             The k-th predator robot find the empty nearest $C(k,g)$ (e.g., greedy strategy)
10:            $C(k,g) = k$
11:        **end if**
12:    **end for**
13:    Find empty cells $C(k,g)$ for the group of communicated predator robots that has the minimum sum of distances of $C(k,g)$ and $A(k,p)$ (e.g., the cooperation strategy).
14:    Determine the guidance points $D(k,g)$ of the univector field $F(k,p)$ from $C(k,g)$ by symmetry $C(k,g)$ through the prey robot's position $g(x,y)$ for all predator robots.
15:    **for** $k = 1$ **to** 8 **do**
16:        $\angle F(k,p) = \angle A(k,p)g - n\alpha$ where $\alpha = \angle A(k,p)D(k,g) - \angle A(k,p)g$
17:        The k-th predator robot moves to next cell based on the value of vector field $F(k,p)$.
18:    **end for**
19:    **if** eight predator robots capture the prey robot **then**
20:        $stop = true$
21:    **else**
22:        **if** a probability of ε **then**
23:            The prey robot moves to an empty neighboring cell at random.
24:        **else**
25:            **if** a probability of q_0 **then**
26:                The prey robot compute the distances between $C(k,g)$ and the nearest $A(k,p)$ and move to a cell $C(k,g)$ that has maximum distance to $A(k,p)(k = 1, \ldots, 8)$.
27:            **end if**
28:        **end if**
29:    **end if**
30: **until** $(stop)$ or $(t < MAX\_ITER)$

At each iteration t of the UFMRN algorithm, each predator robot finds its community in a predefined radius r. If a predator robot does not find any the other predator robots, it finds an empty neighboring cell of the prey robot's position that minimizes its distance. Otherwise, the communicated predator robots cooperate to determine empty neighboring cells $C(k,g)$ that have the minimum sum of distances from them to empty neighboring cells $C(k,g)$. Once all predator robots have determined their cells $C(k,g)$, the guidance points $D(k,g)$ of the univector field $F(k,p)$ are determined by symmetry $C(k,g)$ through the prey robot's position $g(x,y)$. After having the guidance points $D(k,g)$, the values of univector field $F(k,p)$ are determined and then all predator robots move to their neighboring cell based on their $F(k,g)$. For the prey robot, with a probability of ε, it moves to an empty neighboring cell at random. Otherwise, and with a probability of q_0, it moves to an empty neighboring cell that maximizes the distance to the nearest predator robot. The algorithm will end if the prey robot is captured by eight predator robots or after a predefined number of iterations.

Note that, the value of univetor field $F(k, p)$ of the k-th predator robot at the position $p(x, y)$ is the heading angle θ_c of the robot as in equation (3). Therefore, the posture p_s at the position $p(x, y)$ is $p_s = [x, y, F(k, p)] \, T$. To apply this algorithm to the real two-wheel robots, the rotational velocities of the left wheel, ω_L, and the right wheel, ω_R, need to be defined. Since the base width, L, and the radius of the wheels, R, of the robot are known. Based on equations (1) and (2), the velocity vector s as in equation (2) and the robot kinematics $\dot{p}_s$ as in equation (4) are totally determined. A robot moves from the current position to the next position based on its kinematics $\dot{p}_s$.

4 Experiments

We implement the UFMRN algorithm along with proposed strategies to evaluate the efficiency of the proposed method for the pursuit problem. Evaluations are based on the successful capture rate and the sum of moves of eight predator robots for each simulation. In all simulations, we set up an experimental environment with 100x100 grids, all predator robots are randomly initialized at the distinct positions. The coefficient of the univector field $\angle F(p)$ in equation (6) is set $n = 5$. The other parameters are set for all simulations as follows: the radius $r = 10$, $q_0 = 0.9$, $\varepsilon = 0.1$, and $MAX\_ITER = 150$.

To compare the efficiency of the three strategies of the predator robots, we apply these strategies to each strategy of the prey robot as follows:

a) Random strategy

In this case, we carry out 100 simulations. Each simulation is implemented by three steps: Firstly, the predator robots' positions are randomly initialized at the distinct positions. Next, a random path with length of 100 moves of the prey robots is generated. Finally, eight predator robots try to catch the prey robot and the prey robot follows the generated path.

The experimental results show that if eight predator robots use the greedy strategy, the successful capture rate of the predator robots is 78%. If eight predator robots use the cooperation strategy, the successful capture rate of predator robots is 94%. And if eight predator robots use the mixed strategy, the successful capture rate of predator robots is 93%.

Figure 5 shows 50 first simulations. It is clear that the sum of moves of eight predator robots using the mixed strategy is approximately same as using the cooperation strategy. However, the sum of moves of eight predator robots using the cooperation strategy is smaller than that of eight predator robots using the greedy strategy for each simulation.

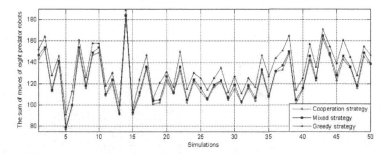

Fig. 5 Comparison of sum of moves of predator robots by using random strategy of the prey robot

b) Greedy strategy

100 simulations were carried out to compare the predator robots' strategies. For each simulation, the predator robots' positions and the prey robot's position are randomly initialized at the distinct positions, eight predator robots try to capture the prey robot based on the designed strategy. In this case, the successful capture rate of the predator robots is 72%, 89% and 87% for the greedy strategy, cooperation strategy, and mixed strategy respectively.

c) ε-greedy strategy

We also carry out 100 simulations. The predator robots' positions and the prey robot's position are randomly initialized at the distinct positions for each simulation. The successful capture rate of the predator robots is 73%, 90% and 90% for the greedy strategy, cooperation strategy, and mixed strategy respectively.

Finally, to evaluate the UFMAP algorithm for the pursuit problem from some special initial positions of eight predator robots and the prey robot, we position eight predator robots from the same side of the prey robot's position. The strategies applied for the predator robots and the prey robot are the cooperation strategy and the greedy strategy respectively. In these simulations, the predator robots almost capture the prey robot. Fig. 6(a) depicts a simulation, where the bold circles are the track of the prey robot and the other trajectories are tracks of the predator robots. The small rectangle shows the position that the prey robot is captured by eight predator robots. Fig. 6(b) depicts another simulation in which tracks and directions of the predator robots are shown at each instant time.

The experimental results show that the pursuit problem can be solved by a simple greedy strategy without requirement for the cooperation strategy. However, if the cooperation strategy or the mixed strategy is applied for multi-predator robots, the successful capture rate of the predator robots is approximately 90%. In addition, if the mixed strategy is used, when the predator robots move nearly to the prey robot position, it will become the cooperation strategy. Therefore, the successful capture rate of the predator robots by these two strategies is approximately same.

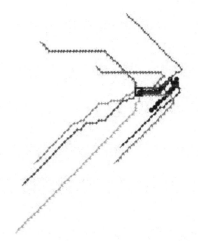

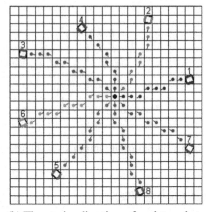

(a) The tracks of predator and prey robots   (b) The moving directions of predator robots

Fig. 6 The capture directions of the predator robots based on greedy and cooperation strategy

5 Conclusion

In this paper, we propose a new algorithm, UFMRN, to solve the pursuit problem based on univector field method for two-wheel robots. In addition, three strategies of the predator robots and three strategies the prey robot are defined to test the proposed approach. For a predefined strategy, at each instant time, each predator robot determines the value of vector field at its position, and then follows the univector field to surround and capture the prey robot. Since univector filed method guarantees mobile robots to reach the target with a desired posture [5]. Therefore, our proposed method always finds the solution of the pursuit problem in the finite moving steps of the predator robots if the environment is large enough. By experimental evaluation, we have successfully shown the application of the UFMRN algorithm to the pursuit problem. However, we have only solved the well-known pursuit problem in an obstacle-free environment. In the future, we plan to extend the proposed approach to the pursuit problem in an environment with obstacles [4].

Acknowledgements. This research was supported by the Basic Science Research Program through the National Research Foundation of Korea (NRF) funded by the Ministry of Education, Science, and Technology (2010-0012609).

References

1. Benda, M., Jagannathan, B., Dodhiawalla, R.: On optimal cooperation of knowledge sources. Technical Report BCS-G2010-28, Boeing AI Center, Boeing Computer Services, Seattle, WA (1986)

2. Campion, G., Bastin, G., D'Andrea-Novel, B.: Structural properties and classification of kinetic and dynamic models of wheeled mobile robots. IEEE Trans. Robot. Automat. 12, 47–62 (1996)
3. Gasser, L., Rouquette, N.F., Hill, R.W., Lieb, J.: Representing and using organizational knowledge in distributed AI systems. In: Gasser, L., Huhns, M.N. (eds.) Distributed Artificial Intelligence, vol. 2, pp. 55–78. Morgan Kaufmann, San Mateo (1989)
4. Kim, Y.J., Kim, J.H.: Univector field navigation method for fast mobile robots in dynamic environment. In: Proceedings of the 2002 FIRA Robot World Congress, Seoul, pp. 165–170 (2002)
5. Kim, J.H., Kim, D.H., Kim, Y.J., Seow, K.T.: Soccer Robotics. Springer, Heidelberg (2004)
6. Korf, R.E.: A simple solution to pursuit games. In: Proceedings of the 11th International Workshop on Distributed Artificial Intelligence, Glen Arbor, Michigan, pp. 183–194 (1992)
7. Reverte, J., Gallego, F., Llorens, F.: Extending Korf's Ideas on the Pursuit Problem. In: International Symposium on Distributed Computing and Artificial Intelligence 2008 - DCAI 2008, 50th edn. AISC, pp. 245–249 (2009), doi: 10.1007/978-3-540-85863-8_29
8. Stephens, L., Merx, M.: Agent organization as an effectors of DAI system performance. In: Proceedings of the Ninth Workshop on Distributed Artificial Intelligence, Eastsound, Washington, pp. 263–292 (1989)

Harvesting Domain-Specific Data Resources for Enhanced After-Sales Intelligence in Car Industry

Jan Werrmann

Abstract. Heterogeneous document landscapes in companies hold knowledge in the form of potential linkage between domain-specific documents of various document systems. To access this (hidden) knowledge, we developed design patterns for an ontology to derive a homogeneous access structure from the heterogeneous document landscape. Like in [19], we describe an *Advanced Ontology-based Information Retrieval System* (AIRS) that includes this ontology to generate retrieval strategies and to find document relationships. In a case study we demonstrate, how the concept of the AIRS can be used to combine the knowledge of different document systems to improve various workshop processes.

1 Introduction

Companies deal with large quantities of different data in a variety of fields. These data result from business process-oriented and domain-specific work procedures and, depending on how they are used, serve as the basis for further processes. The data management systems developed in this way in the past and in various corporate units often operate independently of one another. However, new business processes require this data to be considered as a whole in order to harness knowledge initially hidden that can only be derived by means of semantic grouping and utilize this in various application scenarios.

For example, the following document systems are used in the workshop process in the specialist domains of After-Sales Management at Mercedes-Benz Cars:

- A diagnosis data system in which error code-specific and symptom-based checks are managed for vehicle diagnosis processes,
- A workshop information system in which workshop literature is managed to support vehicle maintenance, repairs and diagnosis,

Jan Werrmann
Daimler AG, GSP Plant 59 Sindelfingen
e-mail: `jan.werrmann@daimler.com`

J. Altmann et al. (Eds.): Advances in Collective Intelligence 2011, AISC 113, pp. 145–167.
springerlink.com   © Springer-Verlag Berlin Heidelberg 2012

- A workshop help system in which documents are managed that provide information about current remedial measures for technical complaints from customers regarding their vehicles,
- Taxonomies for the standardized recording of symptom locations and symptoms in vehicles used by customers, as well as
- Electronic catalogs for replacement parts, work units and flat rates.

However, these systems must be regarded as part of a whole network of systems used in certain areas of the overall workshop process. For example, documents including the workshop information system and the workshop help system are required for the vehicle diagnosis and vehicle repair processes. In the optimized vehicle reception process, symptoms (experienced by customers) should be recorded in a standardized fashion and parts and documents that may be useful for repairing the vehicles should be compiled. Driven by the potential to network service operators[1] with the Mercedes-Benz Cars After-Sales departments that manage the various document systems, the challenge is to bundle the knowledge offered by the systems in order to optimize the workshop process.

Harnessing the full potential for data linking to support the decision-making process would increase the efficiency of the entire workshop process, enabling even better customer service. To achieve this, existing data sources as yet insufficiently linked must be integrated in the decision support process. The assumption of addressing one heterogeneous document landscape with different data gives rise to the following questions:

1. How should the heterogeneous document landscape be modeled –including the potential for linking documents– in order to establish a homogeneous access structure?
2. How are links to be recognized, generated and mapped between (unapparent) semantically correlated documents?
3. How can these links be utilized in an application context?

The approach presented here envisages the establishment of an ontology as a meta level across the heterogeneous document landscape in order to relate documents from different document systems (sources) to each other. In this way, a link can be mapped between semantically correlated documents. These links can be used, for example, to search for documents across all sources.

With a view to the After-Sales department at Mercedes-Benz Cars, this could, for example, produce the following scenario: A symptom experienced by a customer when using his vehicle becomes part of the information requirement[2] in the workshop reception process[3]. During vehicle reception, the task of the service specialists should now be to compile relevant workshop documents as well as any replacement

---

[1] This applied to both branches and authorized workshops.

[2] The need to find documents relevant for repairs from the perspective of the workshop process and with regard to the symptom experienced by the customer.

[3] The reception process is part of the (overall) workshop process.

parts required in order to provide (the customer, for example) with useful information about possible repairs.

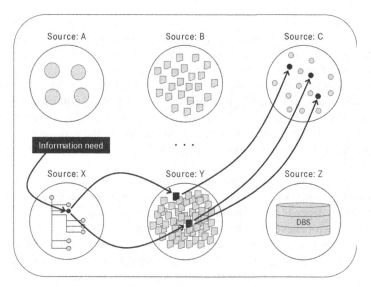

Fig. 1 Heterogeneous document landscape and information need

Figure 1 offers the following insight into exploiting potential for linking documents: Here, a symptom description could be assigned to the taxonomically recorded symptoms, and this symptom could, in turn, be linked with two documents in the workshop help system. Since these workshop help documents could refer to replacement parts in a parts catalog, this means that as soon as a symptom is recognized, replacement parts and informative workshop help documents can be identified for potential repairs.

Thus, the potential of the heterogeneous document landscape can be employed in the relevant application context. For these document links to prove profitable, an ontology that describes the heterogeneous document landscape must be modeled that allows initial tapping of the linking potential offered by the documents. An extended information retrieval system should then be introduced to this ontology in order to control the retrieval process. For example, to traverse relevant data sources in an appropriate order and, in this way, combine fragments of information into semantically-related groups.

In chapter 2, we describe the concept of knowledge presentation and we give design patterns for the ontology modeling process. In chapter 3, we show how this ontology can be used in the context of AIRS. After that we present in chapter 4 a case study and in chapter 5 we give a summary and describe our further research.

2   Knowledge Representation in Context

What is a knowledge representation? In [4], the answer is given by five roles a knowledge representation plays. So, *a knowledge representation* is *a surrogate, a set of ontological commitments, a fragmentary theory of intelligent reasoning, a medium for efficient computation* and *a medium of human expression.* In [1], a *knowledge representation* is defined as *"... the field of concerned with using formal symbols to present a collections of propositions believed by some putative agent.".*

One can argue that knowledge representation involves the mapping of knowledge. Furthermore, it should be possible to use this mapping efficiently in various contexts and to solve certain problems. By *knowledge representation*, we understand a modeling of the heterogeneous document landscape for the purpose of retrieving new, previously unknown information. The model is an ontological description of the domains.

Ontologies can be classified according to degree of formalization and expressivity[4] as well as their domain scope[5]. It must be noted here that integration into these classifications depends on the definition of the term "ontology" itself.

However, we believe that the *application context* should be viewed as a further, significant characteristic feature of ontologies. An **application context** is an application-oriented form of modeling the ontology. In this way, a distinction can made between, among others, the following application contexts:

- A *world of entities* that relate to one another. This allows domain-specific specialist knowledge to be modeled (see, for example, OpenCyc[6]).
- A *lexical knowledge base* consisting of concepts, terms and relations models specialist vocabulary, for example in an intelligent search (such as in [5]) or a concept-based classification (for example in [13]).
- *Modeling of a document world* for ontology-based search. This form of ontology is not based on concepts, but on documents. The documents are modeled as ontological individuals and the document systems are mapped as sources. Existing relations between documents and sources are mapped here as relations between the ontological individuals.

It must be noted that the application contexts may overlap and intermix. Any impact on the modeling of the ontology must then be considered in detail. Thus, an ontology should be made up of three elementary components: **Application context**, **conceptualization** (*classes, individuals* and *semantic relations*) and the **inference**

---

[4] This ranges from informal *lightweight ontologies* through to formal *heavyweight ontologies* (see [7] and [3]).

[5] This ranges from *application ontologies* restricted to one field of application through to *top-level ontologies*, which cover an all-embracing domain (see [9] and [8]).

[6] See [6] and http://www.opencyc.org/, version dated March 28, 2011.

rules[7]. The requirement for conceptualization is fulfilled through classes, individuals and semantic relations. The inference rules embody the degree of formalization and inferences required in order to derive knowledge. In contrast, the application context influences both the conceptualization, the degree of formalization and the explicit requirement for ontological specification.

2.1 Ontology Modeling

In literature one can find various recommendations for ontology building. So, [16] postuleted five phases of ontology engineering (*Indentify Purpose*, *Building the Ontology*, *Evaluation* and *Documentation*). In [2], the five steps *"Emergence of ideas"*, *"Consolidation in Communities"*, *"Formalization"* and *"Axiomatization"* to build an ontology are explained. In [11], five approaches (*"Inspiration"*,*"Induction"*, *"Deduction"*, *"Synthesis"* and *"Collaboration"*) of *ontology-design* are described.

The elements of the ontology created in this article are modeled based on the approach proposed in chapter 2 and, in accordance with this, are divided into three elements: *Application context, conceptualization* and *inference rules*.

2.1.1 The Application Context

To grasp the application context, the domains to be described must be contemplated. The focus is on *top-down* observations of a requirements analysis. These manifest themselves in questions regarding the *application scenario* and the possibility of identifying a suitable model for the domains. Questions regarding how relations are to be derived and whether (and if so, which) factors influence the model must also be answered. Based on the application context introduced in the article, a purely document-based modeling approach is proposed for the ontology. This is realized directly in the conceptualization.

2.1.2 Conceptualization

With regard to the application context and by exploring the heterogeneous document landscape, we have obtained the following insights:

- Documents are managed (or can be located) in document systems (of any type).
- These documents can be assigned attributes/segmented in their document systems.
- Documents from one source are always assigned the same attributes.
- Documents can be connected to other documents (also from the same source) in various ways. *For example: Standardized symptoms and symptom locations*

---

[7] Different approaches also set out the basic elements of ontologies. In [14], it is postulated that in structural terms, an ontology should comprise four elements: *Lexicon* (terminological mapping of domains), *concepts*, *semantic relations* and *rule-based links*. In contrast, the approach presented in this work also considers the *application context* an element of ontologies.

are recorded taxonomically in the After-Sales area of Mercedes-Benz Cars. Approaching taxonomy concepts as documents identifies a taxonomic order (or taxonomic relation) between the documents (taxonomy concepts).

- There are various types of relations: *With regard to the After-Sales area of Mercedes-Benz Cars, standardized symptoms can relate to workshop help documents, for example (is-related-to relation). However, symptoms can also relate to one another (taxonomic relation).*
- Relations and documents can also be valid in certain contexts. *For example: Model series-independent replacement part x is only relevant for a repair if the workshop help document through which it was identified is valid in the context of a specific model series. Otherwise, the vehicle reception service personnel would select replacement parts that do not fit the customer's vehicle.*
- The contextual validities correlate with the information requirement of the person searching (search based on the condition that ... [8]).

It must therefore be possible to map documents and their sources by means of ontological entities. There must also be links between the entities. Furthermore, the attributes of the documents must also be taken into account[9]. This encompasses both the existence of document attributes (at document and source level) and the possibility of assigning attributes to document links. If valid *pathways* can be followed at document level (i.e. multiple document links, see Fig. 1), relations and (optimal) pathways between documents must be defined. To *identify* the potential for linking between documents, relations and (optimal) pathways must also be defined between sources. To achieve this, the connection between document links and source links must be described. Furthermore, validities of documents and relations must be detected by means of modeling or the inference rules.

Documents can have validities, which depends on the context. In contrast to document attributes, context attributes directly influence the validity of the documents. However, the distinction between attribute and context is not trivial. Attributes can provide information about *how* documents (or sources) link to one another. By comparison, contexts indicate *when* (these) links are valid.

Building on this findings, Fig. 2 outlines an ontology mapping the heterogeneous document landscape. Documents are shown (small circles) that are grouped into different document systems (the big circles *Source A*, *Source B*, *Source X*, *Source Y* and *Source Z*). Document links not yet defined in detail are represented by intersections between documents, while source links are represented by intersections between sources. This model does not show the assignment of attributes to documents and sources or context dependencies.

In [19], we address the elements of the ontology further. For example, we describe how document validities and attributes are to be handled during conceptualization.

---

[8] For example: Search for relevant replacement parts based on the condition of a specific model series.

[9] Attributes are assigned, for example, if a document contains field-oriented information. Fields such as columns in an RDBS table or fields of a search index can be regarded as document attributes.

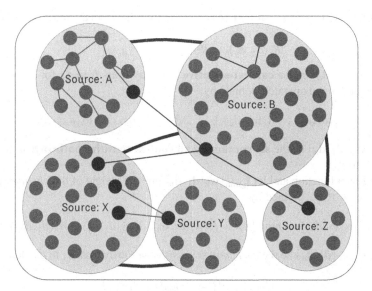

Fig. 2 Ontology representing the heterogeneous document landscape (without attributes and contexts). One can see sources (the clusters *Source A*, *Source B*, *Source X*, *Source Y* and *Source Z*), documents (circles in the clusters) and relations between sources (the thick edges between the clusters) and documents (edges between document circles).

We also explain how the elements of an ontology can be obtained from document systems. A concrete application example is provided in the case study in chapter 4.

2.1.3   Inference Rules

In order to assess relations and identify valid (best) pathways in the ontology, *pathways* must be defined and *weightings* given to valuate these pathways. The objective is to demonstrate how to best describe *optimal pathways through the ontology* and the influence contexts can have on these pathways. The shown formalizations are our recommendation how to define the best context sensitive pathways at document level.

Let $Q_1, Q_2, \ldots, Q_i, \ldots, Q_n$ be sources and let $\mathbf{d_i} \in \mathbf{Q_i}$ be a document of the source $\mathbf{Q_i}$[10].

Let $\mathbf{ka} \in \mathbf{A_K}$ be a context attribute of the context attribute set A_K and let $\mathbf{k}$ be the context of the context attribute *ka*. For a source Q_p:

$$Q_{kp} \subseteq Q_{kap} \subseteq Q_p \tag{1}$$

Let $\mathbf{R(d_p, d_q)}$ be an unnamed binary relation between document d_p and d_q for $d_p \neq d_q$.

---

[10] $d_i \in Q_i$ stands for the *document-of-source* relation between the document d_i and the source Q_i.

Let $R_k(d_p, d_q)$ be the **context-sensitive relation** between d_p and d_q for $d_p \neq d_q$, defined by

$$R_k(d_p, d_q) := R(d_p, d_q)_{d_p \in Q_{kp}, d_q \in Q_{kq}} \tag{2}$$

Let $w_k(d_1, d_n)$ be the **context-sensitive pathway** from d_1 to d_n, for $1 \neq n$ and $\oplus$ as a yet undefined operator, defined by

$$w_k(d_1, d_n) := R_k(d_1, d_2) \oplus \ldots \oplus R_k(d_{n-2}, d_{n-1}) \oplus R_k(d_{n-1}, d_n) \tag{3}$$

Let $W_k(d_1, d_n)$ be the **set containing all context sensitive pathways** $w_k(d_1, d_n)$ from d_1 to d_n.

Let f be a function and let $f_{k\min}(w_k(d_1, d_n))$ be the weight of the *cheapest pathway* from d_1 to d_n, defined by

$$f_{k\min}(w_k(d_1, d_n)) := min(\{x | x = f_k(w_k(d_1, d_n)), \forall w_k(d_1, d_n) \in W_k(d_1, d_n)\}) \tag{4}$$

Let $w_{k\min}(d_1, d_n)$ be the *best context sensitive pathway* from d_1 to d_n, defined by

$$w_{k\min}(d_1, d_n) := f_{k\min}(w_k(d_1, d_n)) \tag{5}$$

The shown formalizations are part of the ones we proposed in [19].

3 Advanced Ontology-Based Information Retrieval System (AIRS)

In line with [12], information retrieval refers to the locating of information (that fulfills a particular requirement) in a (large) unstructured set of data. An unstructured set of data is a collection of documents represented by a document system.

In this context, *locating* means relating an information requirement to a subset within the collection of documents based on a retrieval model and depending on the context. An *information retrieval system* (IRS) is thus a system that supports the *locating* of documents. Heterogeneous document landscapes with different document systems limit the locating of documents with the aid of conventional document retrieval systems since only "relevant", but not "related", documents are found[11].

In our *Advanced Ontology-based Information Retrieval System*, we expand on the traditional approach of IRS by including an ontology component[12] for managing retrieval and document linking. Figure 3 depicts the principal structure of a system of this type and how it works.

As with a conventional IRS, the AIRS comprises two distinct, separate workflows: Indexing and searching. The crucial factor here is *how* ontology component and the IRS interact.

---

[11] In addition, a search can generally only be performed across *one* document system.

[12] Other works also address ontology- or concept-based information retrieval. Approaches are described in [15] and [17], for example.

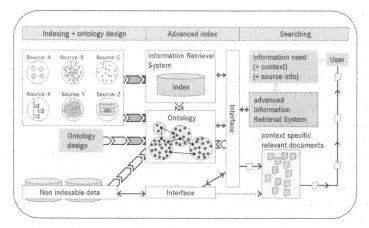

Fig. 3 Structure and functionality of the AIRS. In the part "Indexing + ontology design" (see section 3.1) the document systems will be indexed (if possible) and in the *Ontology design* the ontology will be build. The "Advanced index" will be used in the ontology based search workflow ("Searching", section 3.2).

3.1 AIRS: Workflow Indexing

The indexing process comprises two elements: The individual document systems must be indexed (as far as possible) and an ontology must be created that describes the heterogeneous document landscape. This means that in the *ontology design* (Figure 3, the box, middle left) both sources that can and cannot be indexed must be modeled in the ontology and filled by means of a document processing procedure[13].

This procedure must also include the locating of document links in order to describe relations at ontological level. Another important step in both the indexing and ontology creation processes is the linking of index documents with associated ontology individuals. This is required to enable interaction between the IRS and ontology.

3.2 AIRS: Workflow Search

Figure 3 depicts the following search workflow: A *user* specifies the information he requires by means of a query. He can also choose to define the *context*[14] and *source information*.

To that effect, the response of the AIRS is influenced by the context in that it *only* permits documents considered valid (for the context) to be included in the

---

[13] The procedure should be automated as far as possible, and be at least semi-automated.

[14] Strictly speaking, the system could also determine the context itself, for example by collecting peripheral data.

result set[15]. In contrast, the source information affects the determination of the query strategy.

In this way, *initial and target sources* can be defined by means of the source information. In this case, the AIRS *knows* that the user's information requirement is to be linked with the initial source and that documents from the target source are to be output as the result. Here, the task of the ontology component is to *identify possible links at source level*[16]. The task of the IRS is to provide documents from the initial source. Those documents result from the linking of the information requirement with the initial source documents indexed.

The documents located in this way would then act as the starting point for a further process step[17], which[18] pursues valid, optimal pathways at document level to the target source. This would, in turn, be a task for the ontology component. Finally, the documents *located* in this way are transferred to the user as the result set by the AIRS.

If the *initial source* is specified but the target source is not, the information requirement must be linked with the initial sources (IRS) and documents considered relevant (IRS) could be presented in the result set with *selected* links to other documents (ontology component).

If, however, only the *target source* is specified, the AIRS itself must determine the source with which the information requirement can be linked (IRS) in order to subsequently identify the optimal pathways to the target source (ontology component).

If *neither* the *initial nor the target source* is specified, the sources with which the information requirement can be linked (IRS) are determined by the AIRS (as described for the previous two cases). All documents considered relevant (IRS) are output in the result set with their links to other documents (ontology component).

The next chapter summarizes the tasks of the two elements (IRS and ontology component).

3.3 AIRS: Ontology Component

Locating Potential Links at Source Level

The formal procedure for this task consists of finding the *best context-sensitive pathway* $w_{ka Source} min(Q_p, Q_q)$ from Source Q_p to Source Q_q. For this to be possible, relations between sources must be identified and used to find a pathway. The pathways located in this manner must then be valuated with a weighting function $g_{ka}(w_{ka Source}(Q_1, Q_n))$ and the *best* pathway (if at all available) selected.

---

[15] Accordingly, if the context is not defined, the result set is *not* influenced.

[16] Therefore, the *best* pathway from source X to source Y must have been found.

[17] Which of the documents found are *relevant* for further process steps could be decided by the AIRS or the user himself.

[18] Based on the pathways found at source level by the ontology component.

Pursuing Valid Pathways at Document Level

This task can also be set out in a formal procedure and comprises identifying the *best context-sensitive pathway* $w_{kmin}(d_1, d_n)$ from document d_1 to document d_n or determining the *best context-sensitive pathways* $w_{kmin}(d_p, Q_{kq})$ from document d_p to documents of the source Q_{kq}. This facilitates navigation through the documents on the basis of specific relations and contexts.

Recording Non-Indexable Sources and Documents

Companies do not just have document systems whose documents can be recorded in an IRS. Some documents can *only* be requested via defined interfaces from other systems. Links between these documents and others (documents included in the index of the IRS) must be recorded in the ontology. In this way, the systems to be requested via the interfaces only can be integrated in the AIRS workflow.

***Defining* Relations**

User feedback could be recorded by the AIRS and integrated in the ontology. If, during a document search, a system user *detects* a link between any two documents that has not yet been recorded in the system, this link should be defined as a relation between the two documents and taken into account during subsequent queries. In this case, the task is to label the relation $R(d_p, d_q)$ between the two documents d_p and d_q and specify a weighting $f(R(d_p, d_q))$.

3.4 AIRS: IRS Component

Linking Information Requirements and Documents

The actual task of the IRS is to search across text-rich collections of documents and document attributes. If the query strategy was defined by the AIRS and with the aid of the ontology component, the IRS can be used to search for relevant documents in the indexed data of a specific document system. The *most relevant* of these documents can then be used for further work procedures.

4 Case Study

In order to validate the AIRS approach, a study was conducted to identify the potential for linking various document systems in the After-Sales area of Mercedes-Benz Cars. To do so, the following steps were required:

- Exploring the data sources, including identifying documents, finding attributes, checking documents for context dependencies as well as recognizing and assessing links between documents.
- Generating an index structure relating to the documents identified.

- Modeling the ontological entities.
- Interpreting the recognized document links as relations and mapping these (relations) at ontological level.

4.1 Application Scenarios

Two different application scenarios were considered that play a role in the workshop process:

- In reception, the symptom experienced by the customer must be recorded for downstream work procedures. In the current workflow, the reception personnel interpret the symptom (experienced by the customer) and assign a standardized symptom. To do so, the employee clicks through the symptom taxonomy until he reaches the corresponding symptom node. Symptom assignment should be supported through the generation of proposal lists.
- It would help the vehicle repair process if relevant repair help documents were proposed on the basis of symptoms recorded systematically and replacement parts required for potential repairs were selected.

4.2 Exploring the Data Landscape

As part of the AIRS study, three document systems from the After-Sales area of Mercedes-Benz Cars were examined for potential links: The workshop help system, the symptom taxonomy and a subset of an electronic parts catalog.

Documents and Relations

No problems were experienced when identifying documents in the workshop help system as the structure of the document system meant that the source was already separated in documents. Extracting appropriate attributes and searching the documents for context dependencies proved more difficult. By concentrating on the internal structure of the individual documents, it was possible to select thematic fields that could be interpreted as attributes.

For example, workshop help documents contain "Cause" and "Remedy" sections. Accordingly, the documents were assigned the attributes "Cause" and "Remedy". Due to the fact that they convey information (see chapter 1), both symptom nodes of the symptom taxonomy and replacement parts were regarded as documents. In these documents, properties were then identified as attributes and validities were identified as contextual dependencies. This means, for example, that a symptom node has a language-dependent, textual symptom description and a (unique) ID in the document system. A replacement part has a part text and a part number.

Determining contextual dependencies is particularly challenging: For example, workshop help documents can be assigned to both model series and numerous other validities. Symptom nodes are valid across various model series and replacement parts can be assigned to specific model series. However, other (context-related) validity restrictions may also be identified, devised and, in turn, linked. For this reason, contextual dependencies were not taken into account in the study for the time being.

Links were established between the documents identified in various ways. The following document relations were identified:

- Symptom nodes are integrated in a taxonomy. Accordingly, a taxonomic hyponymy exists between the taxonomy concepts from the superordinate term to the specific term.
- Workshop help documents contains sections that name replacement parts explicitly.
- Workshop help documents include labeled fields with symptom information. These information relates to symptom nodes of the symptom taxonomy.

Based on these findings, we have extracted documents and document links and generated our ontological structure. In table 1 a summary of all extracted documents and relations is given.

Table 1 Count and description of extracted documents and relations.

Type	Description	Count
documents	count of extracted documents	13503
	symptom documents	3297
	workshop help documents	4981
	part documents	5225
relations	count of extracted relations	21916
	taxonomic relations (symptom taxonomy)	3296
	workshop help document to symptom document relations	12535
	workshop help document to part relations	6085

4.3   Transferring the Extracted Documents to an Index Structure

As part of the study, the IRS component of the AIRS was implemented using the open source Apache Solr enterprise search server[19]. The insights obtained from exploring the data landscape were integrated directly in the indexing process. This allowed a complex ETL process[20] to be implemented for copying data from the original systems to the Solr index.

The workshop help documents, the symptom taxonomy and the replacement part descriptions were processed and stored in the Solr index with field-based attributes on the basis of various criteria. The documents identified, as well as their attributes, were assigned to index fields of the Solr documents. This allowed the documents and their attributes to be accessed directly.

---

[19] See http://lucene.apache.org/solr/, version dated April 16, 2011.

[20] ETL stands for Extract, Transform, Load.

4.4 Ontology Modeling

The *Source, Document, Attribute* and *Document Attribute* classes were modeled.
The individuals of the *Source* class are the *Symptom Taxonomy, Workshop Help
System* and *Parts* document systems. Documents in the document systems are represented by individuals of the *Document* class. The document individuals are linked
with their document systems by means of the *document-of* relation.

Attributes are represented by individuals and are linked with the document
or source individuals via an *attribute-of-document* or *attribute-of-source* relation
respectively.

A *hyponym-of* relation was modeled between document individuals of the symptom taxonomy that enables the taxonomic structure of the symptom taxonomy to be
applied to the ontology.

Further document links were mapped in the model with the simple *is-related-to*
relation, which allows document, source and attribute individuals to be connected
to one another.

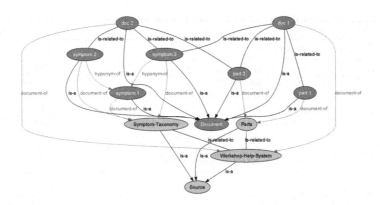

Fig. 4 The ontology contains the **classes** *Source, Document*, the **individuals** *Symptom-Taxonomy,
Parts, Workshop-Help-System* and the **relations** *document-of, hyponym-of* and *is-related-to*. Furthermore, one can see examples for document individuals (*doc_1, doc_2, Symptom_1, Symptom_2,
Symptom_3, part_1* und *part_2*) and document relations (*document-of, hyponym-of* and *is-related-
to*).

Figures 4 and 5 illustrate this approach with some example documents[21]. Figures
6 and 7 shows the ontology derived from the document systems. Figure 6 maps a
subset comprising around 1,000 nodes per document system, while Fig. 7 maps the
entire ontology. The document individuals are shown as nodes and document links
are shown as intersections between the nodes.

---

[21] If contextual dependencies are to be modeled in the ontology in future, this must take place
in the form of *context attributes*.

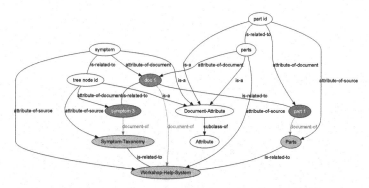

Fig. 5 The **class** *Attribute* is the super class of all attributes. The sub class *Document-Attribute* is a container for all attribute individuals (like *symptom*, *tree_node_id*, *parts* and *part_id*). The attribute individuals are related to the source individuals (*attribute-of-source* relation) and to the document individuals (*attribute-of-document* relation).

4.5 Application in Context

This section explains the AIRS workflow in terms of the two application scenarios. Particular attention will be paid to the roles of the two AIRS components[22].

4.5.1 Structured Symptom Recording

Symptom nodes of the symptom taxonomy are assigned a textual symptom description. This description was recorded as an attributed field in the Solr index when the index was created. In this way, relevant symptoms recorded taxonomically can be searched for when a customer symptom is entered. However, the textual symptom descriptions are only of limited use for document searches since the texts are generally very short[23], giving rise to the familiar problem experienced during text-based information retrieval of discrepancies between concepts and terms[24]. This discrepancy arises because different terms can address the same concept.

In the information retrieval environment, solution strategies call for the generation of conceptual synonym clusters in which terms are combined according to their contextual semantics. The synonym clusters are then combined to form concepts and applied to the retrieval process. This can take place both during indexing and during searches via query expansion technologies[25]. Unfortunately, it may be that the same term addresses various concepts. The ambiguity of terms is one reason for the context dependency of query expansion.

---

[22] The IRS component and the ontology component, see chapter 3.

[23] E.g. "engine rattles".

[24] In literature, this is also termed the word mismatch problem [10].

[25] In fact, further semantic relations are used depending on the query expansion approach. One approach for expanding queries with synonyms and other semantic relations is laid out in [18], for example.

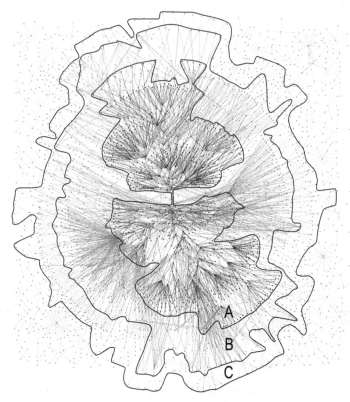

Fig. 6 Part of the ontology, which was derived from the document systems *workshop help system* (B), *symptom taxonomy* (A) and the *parts catalogue* (C). Nodes represent documents and edges relations. One can see the taxonomic structure in the middle of the picture (A).

The findings were integrated in the implementation of the AIRS and, in this way, initial efforts were made to expand the index via synonym lists. The synonym lexicon maintained manually in the After-Sales area of Mercedes-Benz Cars was used for this purpose[26]. The aim was to ensure the domain-specific contextual accuracy of searches. However, the short symptom descriptions meant that it was not possible to locate a symptom recorded taxonomically by means of a textual description. Instead, the symptom (or a synonym defined in the index) had to be named exactly.

Workshop help documents contain large amounts of text and, furthermore, are still directly related to symptom nodes. In a semantic symptom node search, the search space must be extended to these documents. For this purpose, an AIRS workflow for supported symptom taxonomy searches was created in which the IRS component and the ontology component interact. This workflow is shown in Fig. 8. The IRS component is responsible for both of the following work steps:

---

[26] Where possible, every symptom description in the Solr index was supplemented with appropriate synonyms.

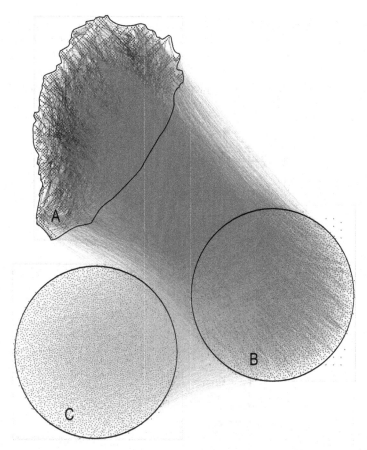

Fig. 7 The whole ontology, which was derived from the document systems. One can see three document cluster, which was extracted from the *workshop help system* (B), the *symptom taxonomy* (A) and from the *parts catalogue* (C).

1. Direct search via the index with the symptom taxonomy components.
2. Indirect search via the index with the workshop help documents. This is only triggered if the direct search is unsuccessful and serves as an intermediate step in which a list is created with workshop help documents.

The ontology component is responsible for the following step:

1. The list generated by the IRS component with workshop help documents is integrated in an ontology search by searching for the corresponding individuals in the ontology created and then localizing the related symptom taxonomy individuals by "following valid pathways at document level" (see 3.3, section: Pursuing valid pathways at document level.) The symptom nodes found then form the result set.

$$w_{k\,min}\left(d_{Workshop-Help-System}, Q_{k\,Symptom-Taxonomy}\right) \qquad (6)$$

from the workshop help document $d_{Workshop-Help-System}$ to symptom taxonomy documents of the source $Q_{k \, Workshop-Help-System}$. Since a baseline procedure was employed in this study, contextual dependencies were not taken into account. Furthermore, the *is-related-to* relation was weighted as "1", which meant that all *context-sensitive pathways* could be assessed as "1" in the model of the supported symptom search.

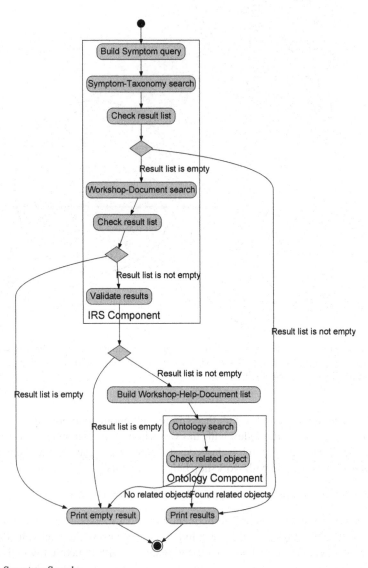

Fig. 8 Symptom Search

In Conclusion

The workflow comprises a direct search for symptom nodes via the index of the symptom taxonomy. If symptom nodes were found, the search terminates. Otherwise, the search strategy is extended, whereby the workshop help documents, which contain large amounts of text, are included in the retrieval process. If the results list is not empty, a search is performed in the ontology for relevant (related) symptom nodes for every workshop help document on the list.

In the study, the approach proclaimed improved the search significantly[27]. Further examinations are to evaluate the extent to which both approaches can be combined and/or the search approach can be extended to cover additional document systems.

4.5.2   Selecting Relevant Documents

The second application scenario requires a list of symptoms recorded taxonomically in order to find workshop help documents and replacement parts connected to the symptoms specified.

In an initial step, the ontology component is tasked with finding the *best context-sensitive pathways*

$$w_{ka\,Source}min\left(Q_{Symptom-Taxonomy}, Q_{Workshop-Help-System}\right) \tag{7}$$

and

$$w_{ka\,Source}min\left(Q_{Symptom-Taxonomy}, Q_{Parts}\right) \tag{8}$$

The pathways are defined in our ontology, resulting in a workflow illustrated in Fig. 9[28]:

1. The first step consists of processing the symptom list sequentially and finding the *best context-sensitive pathways* for every symptom

$$w_{k}min\left(d_{Symptom-Taxonomy}, Q_{kWorkshop-Help-System}\right)$$

   from the symptom document $d_{Symptom-Taxonomy}$ to the workshop help documents of the source $Q_{kWorkshop-Help-System}$. The list created in this way is validated in further steps[29], output as a result and transferred to the second step.
2. In the second step, the workshop document list created is processed sequentially and a search is performed in the ontology for related replacement parts for every workshop help document. In a formal sense, the *best context-sensitive pathways*

$$w_{k}min\left(d_{Workshop-Help-System}, Q_{kParts}\right) \tag{9}$$

   are calculated from the workshop help document $d_{Workshop-Help-System}$ to documents of the source Q_{kParts}).

---

[27] Since above all the first, direct search returned no results in many cases.

[28] The weightings of source pathways were disregarded in the study.

[29] The *best context-sensitive pathways* were weighted as "1" in the study. The validation step was therefore omitted.

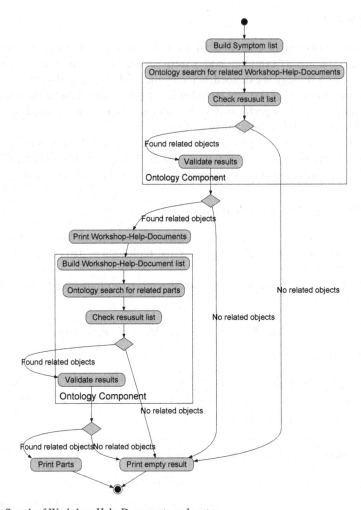

Fig. 9 Search of Workshop-Help-Documents and parts.

In Conclusion

The workflow consists of two-stage navigation through the ontology to find all documents related to a synonym list. Documents of the initial source were defined (symptom list provided) and target sources were specified (workshop help system, parts). The AIRS was tasked with finding pathways at source level and following them at document level. This meant it was possible to specify related workshop help documents and replacement parts for existing symptom lists.

5   Summary and Outlook

This article presented the AIRS approach. It described how the two AIRS components (IRS and ontology components) must interact in order that a cross-document system intelligent search can be implemented and (new) document links can be identified. Theoretical aspects of ontology modeling were also addressed and the need to consider the application context was also explained.

An initial case study showed how the AIRS approach can be integrated in the domains of the Mercedes-Benz Cars workshop process in a beneficial manner. The study showed that document systems (managed by different author processes) can be linked intelligently in order to optimize the entire workshop process by bundling knowledge.

For further steps, it is important to recognize which particular challenges must be overcome for an AIRS to prove profitable. Some questions are raised below that determine what further action will be taken:

- Are sources involved effectively in the information flow and are attributes (to be) assigned to information flows?
- How is this knowledge to be labeled by means of relations and how are relations to be weighted?
- What methods can be used to identify links between documents automatically and what is the quality of these links?
- How exactly are pathways identified in practice?
- What is the distinction between context and attribute?
- What methods can be used to store the ontology?

The following tasks therefore come into question[30]:

Locating, Naming and Weighting Relations between Documents

The challenge here is to evaluate procedures and methods that allow links to be identified (semi-) automatically between (any) documents from (different) document systems. Semantic relations must be assigned names based on these links. In the next step, these relations must be assessed with regard to their expressivity and a weighting function $f(R(d_p, d_q))$ determined[31].

Pathways

Pathways[32] can be *valuated* by means of weighting functions. The challenge is to find the *best* pathways. In the case, weighting functions must be determined for defining pathways through the ontology.

---

[30] More related tasks are described in [19]

[31] It must also be checked whether the relations are context-dependent.

[32] Pathways through the ontology at document, attribute or source level.

AIRS

In terms of actual application, it must be determined how the ontology can be integrated in the AIRS workflow in order to reap benefits and through which exponents of the ontology languages. For this purpose, various approaches to ontology-based information retrieval must be evaluated and an architecture proposal developed.

Consistency

How can the index and ontology be kept consistent? A process must be developed that enables versioning of index documents and the associated modeled individual of the ontology (in terms of the original data).

References

1. Brachman, R., Levesque, H.: Knowledge Representation and Reasoning. M. Kaufmann, Amsterdam (2004)
2. Braun, S., Schmidt, A., Walter, A.: Ontology Maturing: a Collaborative Web 2.0 Approach to Ontology Engineering. In: Proceedings of the WWW Workshop on Social and Collaborative Construction of Structured Knowledge. Banff, Canada (2007)
3. Corcho, O.: Ontology-Based Document Annotation: Trends and Open Research Problems. International Journal on Metadata, Semantics and Ontologies 1(1), 47–57 (2006)
4. Davis, R., Shrobe, H.E., Szolovits, P.: What Is a Knowledge Representation? AI Magazine 14(1), 17–33 (1993)
5. Farhoodi, M., Mahmoudi, M., Bidoki, A.M.Z., Yari, A., Azadnia, M.: Query Expansion Using Persian Ontology Derived from Wikipedia. World Applied Sciences Journal 7(4), 410–417 (2009)
6. Fischer, D.H.: Ein Lehrbeispiel fur eine Ontologie: OpenCyc (in German). Information Wissenschaft und Praxis 55(3), 139–142 (2004)
7. Giunchiglia, F., Zaihrayeu, I.: Lightweight Ontologies. Technical Report DIT-07-071, University of Trento (2007)
8. Guarino, N.: Understanding, Building and Using Ontologies. Int. Journal Human-Computer Studies 45(2/3) (1997)
9. Guarino, N.: Formal Ontology and Information Systems. In: N. Guarino (ed.) Formal Ontology in Information Systems: Proceedings of the 1st International Conference June 6-8, 1998, Trento, Italy, pp. 3–15. IOS Press, Amsterdam (1998)
10. Hoeber, O., Yang, X.D., Yao, Y.: Conceptual Query Expansion. In: P.S. Szczepaniak, J. Kacprzyk, A. Niewiadomski (eds.) AWIC, *Lecture Notes in Computer Science*, vol. 3528, pp. 190–196. Springer (2005)
11. Holsapple, C.W., Joshi, K.D.: A collaborative approach to ontology design. Commun. ACM 45(2), 42–47 (2002)
12. Manning, C.D., Raghavan, P., Schütze, H.: Introduction to Information Retrieval. Cambridge University Press, Cambridge (2008)
13. de Melo, G., Siersdorfer, S.: Multilingual Text Classification Using Ontologies. In: G. Amati, C. Carpineto, G. Romano (eds.) ECIR, *Lecture Notes in Computer Science*, vol. 4425, pp. 541–548. Springer (2007)
14. Sure, Y., Ehrig, M., Studer, R.: Automatische Wissensintegration mit Ontologien (in German). In: K.H. Ulrich Reimer (ed.) Workshop Modellierung fuer Wissensmanagement auf der Modellierung 2006. Institut AIFB, Innsbruck, AT (2006)

15. Tomassen, S.: Research on Ontology-Driven Information Retrieval. In: R. Meersman, Z. Tari, P. Herrero (eds.) On the Move to Meaningful Internet Systems 2006: OTM 2006 Workshops, *Lecture Notes in Computer Science*, vol. 4278, pp. 1460–1468. Springer, Berlin / Heidelberg (2006)
16. Uschold, M., King, M.: Towards a Methodology for Building Ontologies. In: Workshop on Basic Ontological Issues in Knowledge Sharing, held in conjunction with IJCAI-95. Montreal (1995)
17. Vallet, D., Fernandez, M., Castells, P.: An Ontology-Based Information Retrieval Model. In: Proceedings of the European Semantic Web Conference (ESWC). Crete (2005)
18. Voorhees, E.M.: Query Expansion Using Lexical-Semantic Relations. In: W.B. Croft, C.J. van Rijsbergen (eds.) SIGIR, pp. 61–69. ACM/Springer (1994)
19. Werrmann, J.: Modellierung im Kontext: Ontologie-basiertes Information Retrieval (in German). Tech. rep., Department of Mathematics and Computer Science, FernUniversität in Hagen (2011). See http://deposit.fernuni-hagen.de/2769/

Author Index